LIVE SOUND, Theory & Practice
By Mark Amundson

Timeless Communications, Corp.

LIVE SOUND, Theory and Practice

© Timeless Communications, Corp.

First published September 2007 by
Timeless Communications, Corp.
6000 South Eastern Ave. Suite 14-J
Las Vegas, NV 89119
www.plsnbookshelf.com

ISBN: 978-0-9798107-0-1

Exclusively Distributed By
 HAL•LEONARD®
CORPORATION
7777 W. BLUEMOUND RD. P.O. BOX 13819 MILWAUKEE, WI 53213

CONTENTS

CONTENTS

It was nearly a decade ago when I got a phone call from Craig Anderton about some guy by the name of Mark Amundson. I was, at the time, the editor of a musicians' magazine called *Gig*, and Craig was heavily involved with a sister publication called *EQ*. But, being Craig, he was ahead of us all and doing a lot with some chat rooms he ran on the old AOL site. "This guy is contributing pretty heavily to the site and I thought you might be able to use him," he wrote me.

And so began a working relationship that has bridged multiple media types, multiple titles and more than its share of post-trade-show dinners. Mark was then — and still remains — one of the writers I can always count on to give me good stuff without a bunch of spin. As technical editor for *FRONT of HOUSE* magazine, Mark is the person I count on to help ensure we get the details right in everything from product reviews to Showtime listings. He also writes a monthly column and does product reviews in almost every issue of *FOH*.

This is in addition to a day gig as a head applications engineer with a major consumer electronics company, owning and running a local sound and lighting outfit, maintaining a hobby farm and still finding time to play the occasional band gig.

While Mark is a real engineer who understands the physics of sound in ways that most of us never will, holds several patents and has been involved in the development of technology that he can't even talk about (he could tell ya but then he would have to kill ya...), he comes to live audio with the perspective of a musician who has played hundreds of club gigs — usually providing sound as well as playing guitar and keys. The perspective is important, because it explains how Mark can communicate with seasoned pros and green newbies alike, and offer something of value to both of them.

The idea of giving Mark more room to expand his Theory and Practice columns from *FOH* into a book began to coalesce when we started getting requests on a nearly monthly basis for reprints of those columns from instructors teaching live audio courses in colleges and high schools. When we found out that a number of soundco owners were posting his column on the shop wall every month as required reading for their staff, the whole book thing became an obvious next step.

Mark is one of those rare guys who can teach you something without ever making you feel ignorant and whose advice will be not only technically correct, but eminently practical as well. Enjoy the book.

INTRODUCTION

When I started getting into live sound work, I started like many persons do — with a good part of a P.A. that I used in a previous band I played in. But I felt I needed some guidance on the topic, and books and magazines seemed like the logical place to turn, since they had always been a good resource in my previous experience.

The problem was, this was the mid-1980s, and sound systems were barely discussed in the media. Yeah, I got a copy of the original *Audio Cyclopedia*, and read it multiple times like a history book (as it is such). But not much existed in the form of tutorials, and a wide gap was present between a box-mixer and a column of speakers I was dealing with and the pictures of rock concerts showing stacks of big black speakers.

Then a stroke of luck came to me in the form of some old copies of *Recording Engineer/Producer Magazine (RE/P)* found at a ham radio festival while scrounging for P.A. stuff on the cheap. Within each issue of *RE/P*, David Schierman wrote a live sound column about gear brands, configurations and lessons learned from various gigs. It even included an occasional picture or two. For an experience-starved live sound guy like myself, every word was golden — and the ads in *RE/P* influenced me a bit as well.

I say all this because this book is my way passing on things that I have learned in the past couple decades of live sound work. No, I am not some battle-hardened tour veteran from a big national soundco, merely an electrical engineer that doubles as a live sound weekend warrior. Having both professions throughout the '80s, '90s and 2000s has brought me the answers to a ton of questions as to why things have evolved from cobbled-together electronic bits and pieces to purpose-built gear made to handle the ins and outs of club gigging.

Ever since the electronics bug bit me in the mid-1970s, I have strived to find out why engineers make the design choices they do, right down to the last resistor and capacitor. I am not satisfied with "that is how we've always done it," but am driven to seek the real rationale, because the answer can provide the user with some insight on how to best use the designer's products. This has led to a lifelong interest in opening up chassis and finding out what makes these live sound bits and pieces work. By doing this I have gained the respect of many designers, and the disdain of only a couple.

This **LIVE SOUND, Theory & Practice** book is not meant to be a basic introduction into live sound gear operation and how things fit together. There are now a few good books on this topic, and a little hands-on experience is an essential supplement to book smartness. The *Yamaha Sound Reinforcement Handbook* is reasonable recommendation for such rudiments, and I encourage new learners (newbies) to acquire every book and magazine subscription they can avail themselves of. (Especially *FRONT of HOUSE* Magazine, and its free subscription access at www.fohonline. com.)

LIVE SOUND, Theory & Practice is meant to provide practical advice not found almost anywhere else, on real topics important to technicians and engineers in the live sound business. I am not talking about rigging 20-box line array stacks on national tours, but stuff that newbies and national touring crew face as common topics in day-to-day operations. Nothing pisses me off more than reading some high-brow acoustics theory article on a topic that only a couple hundred people on the planet could appreciate. Real live sound engineers swim in a reality of time-pressure, troubleshooting, non-technical individuals (venue management, musicians, patrons, etc., all with their competing demands) and limitations of money.

The 42 chapters of this book are directly taken from my articles in *FOH* magazine since 2003, and constitute a reasonable base of advice for persons in this live sound business. I even find myself going back to the tables and figures to check the data, because there's a lot here to expect anyone to remember it perfectly all the time. It is my fond wish that you will also find the contents worthy of handy reference.

THE PRE-CHAPTER, ELECTRICAL SHOCKS

There is a chart that I keep handy on electrical shocks and the physiological reactions from such shocks. I have several copies of it, and I keep it taped or magnet-ed in easy viewing distance wherever I find myself working. The idea to make it came to me a few years ago when I toiled in the research area of a large heart pacemaker company. I knew we made implantable defibrillators to apply shocks to the heart in order to resuscitate persons with cardiac arrhythmias, so I suspected we had the reference books and Ph.D. brainpower to answer how electrical shocks could hurt people.

My inquires finally led me to a few ancient books of gruesome pedigree: catalogs of Nazi death camp experiments and shock data on American death row convicts volunteering to test the limits of how much electrical shock a human could endure before certain effects occurred. The chart on page 17 summarizes these pieces of science sewed together.

On the chart I list a column that categorizes the AC or DC current needed get differing physiological effects. These effects can range from harmless tingles to muscle contractions to death in various fashions. To make it somewhat meaningful, I did the Ohm's law math at 120 volts to show that skin resistance is everything to surviving electrical shocks. Below the skin, the human body resistance is about 50 ohms most anywhere you can probe with an ohm-meter. But dry skin can be hundreds of thousands of ohms to Mega-ohms, preventing shocks in most cases. There are plenty of stories of old time electricians who were part of the last century's electrification of households, and used their dry leathery fingers to check whether bulb sockets and receptacles were energized or not.

But there is a definite problem when dry hands turn sweaty and the skin resistance drops a lot. From tests with my own hands on an ohm-meter, I can go from 200,000 ohms to below 10,000 ohms just by slightly wetting my fingers to hold the meter probes. This is where tingles turn into serious muscle seizing and cardiac fibrillation, leading to death. That is why I mention in the chart to go to a medical provider after receiving a vigorous shock. I know of a few persons getting a shock earlier in the day, blowing off the event as nothing of consequence, and dying later in the day when their heart jumped into fibrillation.

Fibrillation is a nasty and relatively slow death, and that is why we have Automatic Electronic Defibrillators (AEDs) scattered all over public places. Upon fibrillation, the heart suddenly goes from a single beat per second into a quivering motion that feels odd to the victim at first. But that quivering heart no longer effectively pumps blood, and a brain running out of oxygen from a bloodstream that no longer streams starts a 10 minute clock of death, with consciousness ending within about a minute. If defibrillation is not done within five minutes, the remaining five minutes are now a race against the brain damage that will escalate before defibrillation. So take electrical shocks very seriously. Cardio-Pulmonary Resuscitation (CPR) buys the victim a lot of time, so learn this life saving technique. CPR's chest pushes and breaths do a barely adequate job of getting some oxygenated blood circulation going, but it does do enough to keep you alive, and that's what matters.

Shock Current in milli-Amperes (RMS mA)	Circuit Resistance at 120 Volts AC	Physiological Effects
0.5 to 7mA	240,000 down to 17,000 ohms	**Threshold of Perception:** Large enough to excite skin nerve endings for a tingling sensation. Average thresholds are 1.1 mA for men and 0.7 mA for women.
1 to 6 mA	120,000 down to 20,000 ohms	**Reaction Current:** Sometimes called the Surprise current. Usually an involuntary reaction causing the person to pull away from the contact.
6 to 22 mA	20,000 down to 5,400 ohms	**Let-Go Current:** This is the threshold where the person can voluntarily withdraw from the shock current source. Nerves and muscles are vigorously stimulated, eventually resulting in pain and fatigue. Average let-go thresholds are 16 mA for men and 10.5 mA for women. Seek medical attention.
15 mA and above	8,000 ohms and below	**Muscular Inhibition:** Respiratory paralysis, pain and fatigue through strong involuntary contractions of muscles and stimulation of nerves. Asphyxiation may occur if current is not interrupted.
60 mA to 5 A	2,000 down to 24 ohms	**Ventricular Fibrillation:** Shock current large enough to desynchronize the normal electrical activity in the heart muscle. Effective pumping action ceases, even after shock cessation. Defibrillation (single pulse shock) is needed or death occurs.
1 A and above	120 ohms and below	**Myocardial Contraction:** The entire heart muscle contracts. Burns and tissue damage via heating may occur with prolonged exposure. Muscle detachment from bones possible. Heart may automatically restart after shock cessation.

SPEAKERS AND AMPLIFIERS

Speakers and ampli-
fiers are the heart of any sound
system, even though I actually designate
the priority of such purchases as very low. There
are always difficult tradeoffs to be had, and the
next chapters will discuss a number of these aspects.
The picture below is a reminder of where we have come
from, as it pertains to live sound reinforcement. This circa
1921 photo of a primitive microphone (reflector), amplifier and
loud speaking apparatus was one the first "public address" systems in which a person could address a thousand people and
still speak at a moderate volume. Most of these systems were
built one at time to custom requirements, and remained that
way up to World War II. After the war, names like, DuKane,
Bogen, Rauland, Stromberg-Carlson and Altec provided
standard P.A. components, but mostly for speeches and
modest volume singing needs. Not until the late
1960s did the demands of loud rock 'n' roll music change the requirements of the music equipment manufacturers.

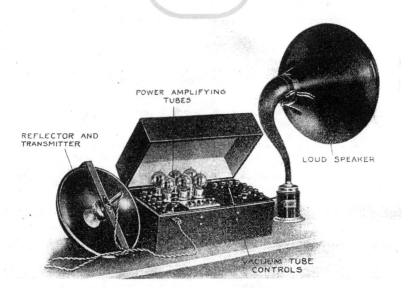

POWER AMPLIFYING
TUBES

REFLECTOR AND
TRANSMITTER

LOUD SPEAKER

VACUUM TUBE
CONTROLS

I learned about speaker sensitivity and power handling by example long ago. Back in the late '70s, I installed my $29.99 below-dash cassette player from Radio Shack, and I needed a couple of 6-inch by 9-inch speakers for the back dash of my trusty '71 Chevy Nova. Being a high school senior saving most every penny for my future electrical engineer school tuition, the then popular 30-watt co-axial 6-inch by 9-inch speakers were both pricey and over-spec for my lo-fi 4-watts per channel at 4 ohms player. So I scrounged up two 6-inch by 9-inch speakers from a junker Magnavox console TV set.

The speakers were stamped on the AlNiCo magnet assembly with 5 ohms and 5-watts, and looked to be a better match for my 4-watt amplifiers. My intuition told me that these speakers should be more efficient (sensitive), since they would give all they had at about 5 watts, and I suspected the 30-watt version would not necessarily give me six times the acoustic power — even if I purchased a 30-watt power booster unit. The auditionable result confirmed my suspicions, and I still had plenty thump coming from the back dash. So much so that the image in my rear view mirror blurred to every kick drum and bass note played through the system.

FAST FORWARD TO THE PRESENT

From frugal student to soundman curmudgeon, I still get incensed about speaker cabinet specsmanship when I look at a broad range of sound reinforcement speakers and see the sensitivity ratings. In general, I see low to upper 90s (dB SPL at 1 watt and 1 meter) for the MI offerings, and upper 90s to mid 100s for the touring rigs. And just when Joe Garage Band needs every decibel possible for his limited and affordable watts, he gets underserved by the same speaker manufacturing industry that spends lots of ad dollars vying for his attention. You would think the manufacturers would shrink the frequency response, or the cost/quality of the rat fur covering, in order to put the extra pennies into more efficient drivers.

I know it is not all black and white on MI to Touring grades when it comes speaker sensitivity. Obviously, long throw cabinets and horn loading will add a couple dB in the calculation. But even when you level the field on enclosures with direct radiating low-frequency drivers and wide coverage angles there still is a difference. After all, the touring guys can afford supersized amplifiers and can tolerate a little less speaker cabinet efficiency compared to the weekend warrior musicians and the ankle-biting club-dwelling soundco's.

THE PROSECUTION RESTS

Taking the other side of the argument, I know that specsmanship and driver design trades all affect the final published speaker sensitivity. Even when I shop for speaker cabinets, I give a bit of leeway to the poor cabinet designer who accidentally found a dB or two of dip in their frequency response curves at 1kHz that manifests itself as lower sensitivity number when comparing apples to apples. Some manufacturers are just prickly enough about the sensitivity specification that they will apply a wide bandwidth filter to the raw frequency response numbers to bump up another decibel in claimed sensitivity. As an enlightened consumer, you have to put yourself into the speaker cabinet designer's shoes and imagine the pressure the product marketing manager applies to make the preliminary product specification goals.

WHAT IS IT REALLY ALL ABOUT?

When you get down to it, the speaker comparision game comes down to quality and quantity. When you ask a group of live sound engineers for speaker recommendations, they typically gravitate to brands of known quality and reliability. The history of the pro-sound speaker manufacturing is littered with sound companies that started trying to find bits and pieces of drivers, cabinets, crossovers, hardware and cover materials — anything that would likely survive the bumps of the road and sound decent. Many of these sound company founders who succeeded in assembling the "good stuff" into road worthy enclosures are today's pro-sound speaker manufacturers.

But today we have a pro-sound-speaker buying populace that has reached sufficient critical mass that several dozens of manufacturers all over the globe offer reasonably complete lines of speaker cabinet choices. Each vendor has done plenty of market research and knows what configurations sell how many units per year. The result is many similar looking speakers with different logo-plates. As buyers, the quality has risen to the point that the lesser-known brands field road-worthy gear. So are we back to specsmanship?

WHERE'S THE BEEF (LOUDNESS)?

Turning to the quantity side of the proposition, available Sound Pressure Level (SPL) is usually rated in decibels at a standard 1-meter reference standoff. Most of the MI offerings range in the 120 to 130 dB SPL range on a continuous basis. Most manufacturers will offer both continuous and peak values, some will only offer the peak. Besides performing the raw SPL measurements, both manufacturers and buyers should know that the final SPL can be calculated pretty closely by combining the sensitivity and the power handling numbers.

For example, a dual 18-inch subwoofer cabinet could contain two drivers each with 400-watt continuous power handling ratings. If the pair in a ported, front-baffled box provides a sensitivity of 101 dB SPL at 1 watt and 1 meter, then providing an 800 watt signal in the center of their frequency response should provide about 130 dB SPL continuous at 1 meter. The reason for this math is the fact that 10 times the logarithmic value of 800 is 29, or 29 decibel watts. A reputable manufacturer may give both the 130 dB SPL continuous and a 136 dB SPL peak rating, but a less scrupulous vendor may just give the 136 dB peak, hoping that you space out about the peak suffix and compare 136 to other vendor's continuous rating numbers.

How can you get 136 dB peak SPL at 1 meter, you ask? In the previous example, our 400 watt drivers may indeed have 800 watt program power ratings and 1600 watt instantaneous (peak) power ratings. The same example subwoofer may then advertise a 3200 watt peak power capability, and ten times the logarithm of 3200 is 35 — 101 plus 35 gets you 136 dB peak SPL.

MORE POWER, SCOTTY

This brings us back to what speaker cabinet manufacturers have traditionally hyped: power ratings. If you are a speaker designer, you know that approximately 80 % of the amplifier watts delivered to a low-frequency driver is lost in heating the voice coil resistance. The result is speakers that are tested with pink noise signals and pumped up until the voice coil wires reach the fusing (melting) point. This becomes the thermal limit of the voice coil in watts, and backing off a touch to survive the long-term testing gives us the continuous (RMS) ratings.

But in reality, music signals are not short-term continuous test signals. To reach an equivalent voice coil thermal power input using music signals, typically a doubling of the music power levels (+3 dB) is arrived at. This is defined as the program power rating. This is around the maximum power you want to size an audio power amplifier for the speaker. If you play the sensitivity and power numbers game, use the program levels to best depict the acoustic capabilities of your system. In the example, the drivers are rated at 800 watts program and together create 133 dB SPL at 1 meter given an amplifier that delivers the desired 1600 watts.

Peak power ratings are more of a driver physics proposition than a power issue. Most driver designers will aim for a +6 dB over continuous power rating as the peak power capability. This limit is what the driver can mechanically (acoustically) perform before the driver component damage occurs. This damage scenario could include voice coil misalignment and popping out of the motor assembly, the coil banging against the back of the magnet or magnetic motor assembly, folding of the cone or tearing of the cone surround or spider. With today's ultra-high-power amplifiers, sound companies are beginning to see more driver mechanical damage signs than fused voice coil windings.

One other point to make is that driver design has progressed so much over the recent years that thermal design of voice coils and coil formers have improved to the point that secondary thermal issues have become primary considerations. These come about when the heat dissipated by the voice coil now efficiently transfers to the nearest thermal mass, the magnetic motor assembly, and heats the magnets. Unfortunately, magnetic materials have a bad property of losing magnetic field strength if they surpass a certain temperature (called Curie temperature). This results in speakers that actually get quieter as the concert goes on, and may even spiral into driver thermal destruction if the person at the console continues to crank it up to compensate.

MY SOAPBOX

The point of this blabbering about specifications and physics is that I want speaker buyers — especially entry-level buyers — to really pay attention to sensitivity ratings when doing apples-to-apples product comparisons. Next, I want speaker manufacturers to stop foisting these trashy, flashy entry-level cabinets with bad specs onto consumers. If I choose a cabinet with 3 dB less sensitivity than the others, that means I need 3 dB (twice) more power to get the same loudness. Sensitivity improvement is not necessarily a vice of the "rich and touring," but a design aspect that can be brought into cost-effective products as well.

In numbers, this means that single driver boxes should be able meet or exceed 98 dB SPL at 1 watt, 1 meter. I feel put off by long-time and reputable manufacturers offering 95 to 97 dB SPL designs and even trying to market them in a professional context. I am even more insulted by seeing these "DJ-market" crumble boxes sporting a piezo tweeter or two with 92 dB SPL sensitivities. If I wanted that kind of power efficiency, I know where the hi-fi store is. Yes, quality of sound is important and should command premium dollars for premium performance, but the business of sound reinforcement is also getting the job done efficiently.

2 PRACTICAL LINE ARRAYS

July 2003

Everybody seems to getting into line array mania, or at least that is how it seems judging by all the marketing and press. This chapter shall dive into the whys and why nots of using line array speaker cabinets, and discuss the various types of arrays.

WHY NOT A LINE ARRAY?

Listeners benefit from line arrays when the arrays become nearly as large as the wavelengths of sound they need to reproduce. That means you should be prepared to have at least about 6 to 40 feet of speaker height clearance, with the lowest speaker cabinet many feet above the audience heads. This tends to exclude low-ceilinged venues. Also, short audience depth venues (less than 60 feet) do not benefit greatly from line arrays, as the arrays designs tend to be geared to direct tight vertical beamwidths into the audience spaces. Shorter venues need broader vertical beamwidth coverage. Beamwidth can be described as the degrees of arc a sound wave spreads (vertically and horizontally), where the power is greater than a fourth of the maximum level.

ADVANTAGES OF LINE ARRAYS

Other than these few small venue dimension exceptions, line array speaker systems have a couple advantages over traditional speaker cluster arrays. The first advantage is that the coupling of a vertical line array can focus the sound to the patrons at the exclusion of everything else. This means less acoustic energy is bouncing off the ceiling and sidewalls, and more is focused into the seating areas. This also means less reverberation and more intelligibility in the mix. Also, this increased directionality means less sound on the performance stage.

Another line array system advantage is the cylindrical expansion of the sound waves. Traditional speakers are treated as point source emanations of sound, with sound wave expansion in both vertical and horizontal directions. With line arrays, the expansion is mostly horizontal in direction, which results in a −3 dB per distance double loss of sound pressure level (SPL) in the near field. Point source speaker systems suffer from the spherical −6 dB per distance double loss, which means you gotta blow the ears off the closest audience members to get reasonable SPLs to the back rows. By taking advantage of the lower-loss near field effect of line arrays you can lower the cabinet SPLs, use less audio power and gain clarity of the sources. One reason for the increased clarity is that reverberations tend to exit the near field and resume the −6 dB loss in the far field region, taking the surface bounce and return while still dropping per the –6 dB rate. This means reflections are greatly reduced compared to the forward SPL of line array.

LINE ARRAY THEORY

The nature of line array speaker systems is that sound sources can combine additively if the spacing is a half-wavelength or less between each adjacent source. To get a feel of wavelengths, one must grasp the concept of the nominal speed of sound (1100 feet per second) and the frequency of the sound waves in the audio band. At 200 Hz, the wavelength of a sound wave is about 6 feet. At 2 kHz, the wavelength is about 7 inches. At 16 kHz, the wavelength is about 0.9 inches. As you can see, these dimensions vary greatly, and speaker designers must use non-traditional methods to reproduce the higher frequencies in order to maintain the multi-speaker coupling and near field effect.

In **Figure 2-1**, a minimalist line array system is shown, with four cabinets in a uniform array. If the cabinet (box) dimension "b" is about a wavelength apart, a 60 degree beamwidth results. If the "b" dimension is about four wavelengths, the cabinets will couple to a 15 degree vertical beamwidth. To keep line array cabinets tightly coupled over mid-bass to high frequency regions, most use a tri-amp configuration with the three driver sections set up to use most of the vertical space in the cabinets, while still providing 90 degree or wider horizontal beamwidths to minimize the number line array stacks to be flown.

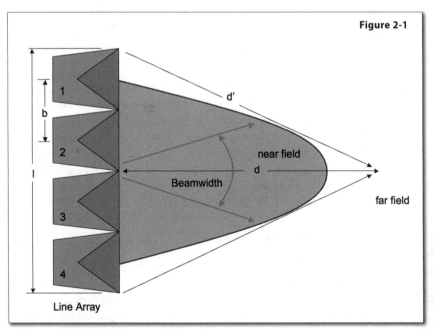

Figure 2-1

Line Array

Now I know the burning question in your minds is, "Where is the boundary between near field and far field in line arrays?" The answer is not perfect, since the boundary is dependant on the widely varying wavelength or lambda (λ) of sound waves and the total height (h) of the line array assembled. Per **Figure 2-1**, when length d' minus a quarter wavelength equals distance "d" is where the boundary exists. Putting it in terms of array height, the equation $d = h^2/2\lambda$ approximates this boundary. In layman's terms: The taller the array, the longer the near field effect, and the better lower frequencies stay in proportion with the mids and highs.

ARRAY TYPES

Per **Figure 2-2**, three major line array types are used commonly. They are the uniform "flat" array, the constant splay array and progressive splay (or "J") array. The uniform array tends to be used in two to eight box configurations in smaller venues, or in large numbered arrays for flat ground outdoor concerts. As the array size gets eight wavelengths or greater a very narrow vertical beamwidth results. This well-coupled, high-powered array, when slightly tilted downward, can cover several acres of music fans with great sound fidelity.

Constant splay arrays are popular in arena settings with the audience located partially on the main floor and in multiple balconies (raked seating). By increasing the tilt (pitch) angle evenly on each of the individual speaker cabinets the result widens out the vertical beamwidth to permit a more even vertical spread of sound, while preserving most of the near

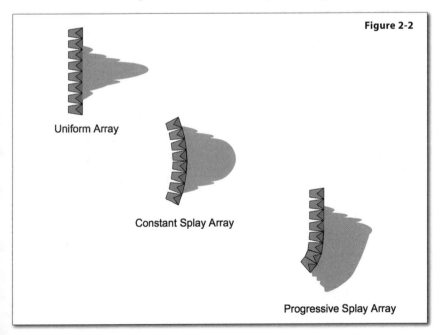

Figure 2-2

Uniform Array

Constant Splay Array

Progressive Splay Array

field effect. This array type is popular in sports arenas where the speakers are flown high with little expected coverage to areas below the array.

The progressive splay line array is the currently the most popular array system used in concert halls and arenas with lots of main floor seating plus the normal raked seating. It is a blend of the previous two array types by starting with a uniform array and progressively pitching the lower cabinets in increasing degrees of splay. This places most of the total acoustic power to the more distant elevated seating while providing lesser but near equal loudness to the main floor. The last one or two lower array cabinets tend to be optimized for greater individual beamwidth as "downfill" cabinets to get to the front row seating.

VERY LOW FREQUENCIES

The lowest octaves are ignored in most line arrays, simply because the near field effect can not preserved in the array, and also because the nearly omni-directional characteristics of sub-woofer cabinets make it better to couple them with the stage or show floor to preserve most of the "half-space" acoustic gain (+3 dB SPL). In larger venues where large amounts of sub-woofer energy can not spread well, flown arrays of sub-woofer cabinets are used to supplement the ground stack subwoofers.

D amping factor may not be the best method to address amplifier to speaker connections, but the parameter reflects the quality of materials and the electrical performance that results in measurable audio fidelity. Consider this my roundabout way of getting on my soapbox to use good cabling and short-as-possible cable runs.

Damping factor is traditionally a way of defining the ratio of load impedance to the source impedance. A larger number means a speaker cone is well-controlled and permits "tight" sounding bass notes — instead of "flabby" sounds that remind you of plucking a barely taut rubber band. Most solid-state audio power amplifiers for live sound applications can readily approach or exceed a damping factor of 1000 for a specification. Given an 8 ohm nominal speaker load, the effective amplifier output impedance of 8 milliohms can be had even if the schematic of the amplifier components have hundreds of milliohms of path impedance. The reason for this is that the hardware output impedance is divided by the negative feedback loop gain of the audio power amplifier. The feedback checks the input reference signal for any sign of output signal mis-tracking, and is corrected by the amplifier circuits — to provide a result that is an effectively lower output impedance.

A REALITY CHECK

In reality, there are plenty of other source impedance aspects that need to be accounted for in the damping factor computation. If the amplifier output jack, speaker cable, cable connectors, speaker input jack and speaker cabinet wiring add a half-ohm of source impedance in series with the amplifier output impedance — then the damping factor of 1000 gets reduced to about 16. But there is good news and bad news in these computations. The good news is that humans generally can only detect flabbiness if the damping factor gets below 10. Some golden-eared engineers can claim detection up to a damping factor of 20. So, shooting for a not-lower-than-10 and better-than-20 system goal is reasonable in choosing speaker cable types and factoring connector losses. The 10 to 20 value range of Damping Factor results in a millisecond or two of damping time constant. With a typical 100 Hz bass note extending its cycle over 10 milliseconds, you can get a feel for how much acoustic waveform control you have.

Another reality check is that historically we have been brought up with amplifier-speaker combinations that have had horrible damping factors. Vacuum tube power amplifiers typically have output transformer windings with 10 % or more of the load impedance expressed in the speaker winding providing damping factors in the 2 to 10 range without other aspects taken into account. Many live sound — and just about all hi-fi — speakers include passive crossover networks with inductors in the low-frequency driver path. These inductors block high frequency audio content, but also add many hundreds of milliohms into the signal path, further degrading the damping factor.

VOICE COIL IMPACTS

The final insult to most golden-eared sound engineers is that if you consider voice coil parasitics, also known as the non-contributing impedances, to speaker operation, the damping factor would never get above one. About 80 % of the voice coil impedance is the wire winding resistance that does nothing but produce heat. The remaining voice coil inductance does the work of moving the cone by storing energy in the magnetic circuit and working against the permanent magnet motor of the speaker basket. For example, a typical 4-ohm low-frequency driver (woofer) has about 3.2 ohms of winding resistance. The remaining impedance is the voice coil inductor containing an inductance (reactive impedance) plus counter-EMF (Electro-Motive Force) voltage that the amplifier must dissipate through its low impedance circuit paths for tight bass response. By moving the 3.2 ohms from load impedance to source impedance you can see that we live in a poorly damped speaker reality. However, by keeping our wiring and connection losses low, we gain every chance to control speakers effectively.

PULL OUT THE ABACUS

Lets ignore the dismal news of voice coil resistive losses, and run the numbers on a common example for an amplifier-speaker connection. Using a modest 20 milliohm output impedance power amplifier driving a 50-foot length of a 12-gauge NL4 Speakon patch cable to a 4 ohm subwoofer speaker cabinet, we can compute a nominal damping factor. The first item is to find all the resistive losses between the amplifier circuits and the voice coil. Some will be readily available, some will be found by heuristics (your best guess). NL4 Speakon jack-to-plug connections are rated by Neutrik to be about 3 milliohms per connection fresh from the factory. I like to round these up to 10 milliohms per conductor connection to take into account wire/socket oxidation and connector contact oxidation. With two interconnects, that adds another 20 milliohms. Next, look at **Table 3-1** for 12-gauge wire and note the 1.650 milliohms per foot specification. Since a 50-foot patch cable has to complete the circuit using two 12-gauge wires, a total of 100 feet, or 165 milliohms, are tallied into the source impedance.

Wire Gauge	Ohms/foot
16	0.004172
14	0.002624
12	0.001650
10	0.001038

Table 3-1

Adding up the source impedance gives us 20+20+165 or 205 milliohms. Tacking on another 5 milliohms for inside the speaker cabinet wiring totals to 210 milliohms, or 0.210 ohms estimated source impedance. Given the nominal load impedance of 4 ohms, the computed damping factor is 4/0.210 or about 19. Now, if we force a nasty situation by substituting 16-gauge cable for the 12-gauge cable (in other words, getting cheap), a 100 feet of 16-gauge is 417.2 milliohms. Taking this value and adding the 45 milliohms of other source impedance losses gives us a new damping actor of about 8.6. This value is below the 10 criteria and swapping between cables would likely be detected in blind A/B tests.

ELECTRIC HEAT

Another aspect of interconnect choices is the resistive losses that begin to be sizable power losses, especially in professional live sound reinforcement using kilowatt level power amplifiers. In the above example, the 50 feet of 12-gauge speaker cable would provide about a 4 % drop of power into a 4-ohm speaker load. If that load is powered by a 2100 watt amplifier, more than 80 watts are lost as heat in the copper wire. In the 16-gauge example, about a 10 % loss is expected, and about 220 watts of the 2100 watts peak signal is left in the copper wires. While these losses may not be easily heard as a loudness loss, the change in damping factor and slight cable warmth may be detected.

PARTING THOUGHTS

It needs to be repeated that short lengths of speaker cables with high-current-capable interconnects remain your best bet to not "hear" your cabling. While I listed example numbers in near-perfect conditions, all it takes is a little bit of contact corrosion or a loose connector termination to make things sound terrible. If possible, choose Speakons or Elco connections over banana and phone connector options to minimize contact resistive losses (count those milliohms). Also invest the time and materials to wipe the connector contacts with a light coating of contact cleaner. I use cotton swabs and the Caig Labs two step process of "De-Oxit" and "Preserve-It" on cabling periodically. Old soundmen will know these chemicals as Cramolin Red and Blue, and they have been around since the birth of the vacuum tube.

When you say the phrase "bi-amped," most newbie sound persons are in awe that you have such a system, and they think bi-amping can cure the common cold — or at least all vestiges of bad sound and feedback. What I want to do in this column is introduce the bi-amping method, and show through examples how you can do it correctly without a lot of information like speaker processor presets and driver response curves.

Looking at the flip-side, using an existing speaker cabinet's passive crossover network solves all your problems as the system becomes a one-amp, one speaker, plug-and-play operation. For a majority of users, the compromises of a passive crossover network inside the cabinet are acceptable. The compromises are: losses in damping factor and efficiency due to passive component impedances; a blurring of fidelity at the crossover frequency point due to non-phase aligned high and low frequency (HF and LF) drivers; bumps in the frequency response due to a non-optimal crossover frequency point and due to keeping the passive crossover component count low; lastly, the imprecision of the HF driver efficiency matching and the necessary protection circuits as low frequency amplifier clipping can produce harmonics that could damage the HF driver.

Bi-amping may resolve most of these problems, but if not done correctly it may produce more harm than good. In the next section, I'll show a basic example and extend it to bi-amping with an analog crossover, then take it to the next step by using a speaker processor. **Figure 4-1** shows the basic schematic of a bi-amp speaker system.

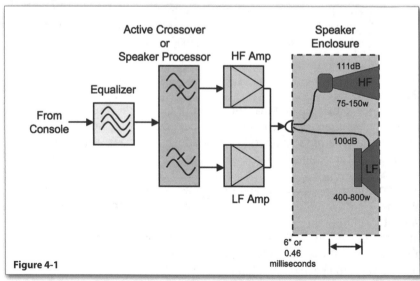

Figure 4-1

A WORKING EXAMPLE

Let's say that we have inherited someone's homebrew (a.k.a. "proprietary") monitor wedge loaded with an Eminence Delta Pro-12 LF driver and a Peavey 22XT HF driver for bi-amping. The first thing to do would be to ensure a suitable four-wire connector is in place (such as a Neutrik NL4 Speakon), with completely separate runs from the driver connections to the connector. Making sure the +1/-1 contacts go to the LF driver, and the +2/-2 contacts go to the HF driver reduces the risk of blowing the HF if you connect to a two-wire passive amplification system. Another optional protection would be to add a non-polarized 10-microfarad capacitor (50-volt minimum rating) in series on the +2 wire, to prevent accidental low-frequency energy from getting to the HF driver. While some are concerned with non-DC coupling the HF driver under SPL conditions, the choice to protect or not is your educated risk.

Bi-amping is more than two amps to two drivers, and even more than adding an active crossover unit before the amplifiers. First of all, the HF amplifier can be one-half to one-fourth the RMS power rating of the LF amplifier, so bi-amping is conducive to running pairs of bi-amped speakers with two stereo power amplifiers, each sized for their types of drivers. In the example, the 22XT HF driver can handle 75 watts continuous (RMS) from 1.5 kHz and up, with the Delta Pro-12 handling 400 watts continuous above 100 Hz. With both drivers having an 8-ohm impedance, amplifier sizing can be chosen from the RMS rating to twice the RMS rating, or about the program rating of the drivers. This means the LF amplifier should be about 400 to 800 watts at 8ohms, and the HF amplifier should be 75 to 150 watts at 8ohms.

CROSSOVER POINT

Without a preconceived notion of the proper crossover frequency point, there are some rules of thumb to use. HF drivers with 3 or 4-inch diaphragms should work fine at a 1.5 kHz point, and HF drivers with 1 or 2-inch diaphragms should work well at the 2.5 kHz point (such as the 22XT). But before you dial 'er up and kick on the power amps, there is one more item to consider: The efficiencies of LF and HF drivers vary quite a bit, with the horn loaded HF usually many dB higher than the LF. So when doing first time trials, back off on the analog crossover gain by around 6 dB to the HF section, relative to the LF section.

For you seat-of-the-pants operators, this is all you need to know. By using your most precise measuring instrument (your ears), you should be able to tweak in the HF and LF gains to taste, and trim the crossover frequency points within a few hundred Hertz of the "thumb" values. For those of you who would like a more scholarly view of bi-amp setup, more information needs to be acquired or measured.

INFORMATION OVERLOAD

A handy set of specifications to have would be the efficiency ratings of each of the drivers. In our working example, the Pro-12 LF driver is rated at 100 dB SPL at 1 watt, 1 meter, with the 22XT rated at 111 dB SPL at 1 watt, 1 meter. Thus, this 11dB efficiency difference could be factored into analog crossover gain settings or in a digital speaker processor. If this data is not available, you can measure similar results using a Real-Time Analyzer (RTA) and a pink noise source. At moderate volume levels an RTA fed by an omni-directional measurement mic can show the approximate frequency response curve when the crossover is fed a pink noise signal. Pink noise is the hissing signal source that has equal energy in each octave of the audio bands. Most RTAs use a display screen or a matrix of LEDs with a gain control to depict the frequency response.

Measuring and tweaking the bi-amp system can be done by placing the LF driver gain at unity (0 dB) and adjusting the HF driver gain so that a reasonably flat response is in the area around the crossover frequency. Moving the crossover frequency point a little bit may further squelch some bumps and unevenness in response curve. When using external analog crossovers, beware of using a "CD Correction" button to help in the driver matching. This feature is for placing a high-frequency shelving filter boost to correct for losses when using constant directivity (CD) horns on the HF driver. Use this feature if it fits well, but avoid it if the one-size-fits-all setting becomes too shrill in the presence (2 kHz to 8 kHz) and high (8 kHz to 20 kHz) frequency bands. By using a graphic equalizer before the analog crossover, you may be able to tune-in a more optimal response.

GOING DIGITAL

Jumping into the world of digital speaker processors not only lets you adjust driver gains and crossover frequencies, but additional features like signal limiting, time alignment and parametric equalization are added. By employing the features in the digital realm, your sound fidelity may improve from good to gonzo! And if you are lucky, and use popular manufactured speaker cabinets, some speaker processors may include factory presets optimized for your system.

With the crossover frequency and driver gains, the previous setup information still applies. Using the limiter setup menus, you may use your amplifier's full power sensitivity input signal levels as a guide to limit the driver excursion just as, or slightly before, the amplifier hits its clipping/limiting threshold. More advanced speaker processors are offering compression or overshoot limiting features to keep your driver on the safer side when the speakers are being pushed to the maximum. Low frequency drivers can have softer limits or a more gradual entry into compression by a couple dB, with HF drivers needing none to maybe 1 dB of hard compression before brickwall limiting.

DELAY ALIGNMENT

With most passive and analog crossover bi-amp systems, frequencies near the crossover point come out of both drivers. But most speaker cabinets have the HF driver behind the horn assembly, and this does not line up with the voice coil of the LF driver. Those few inches of HF to LF voice coil misalignment can create some phase cancellation and blurring of the music in that frequency band. By using the digital signal processor, a fraction of a millisecond of LF delay can be added to bring the drivers phasing into alignment. For example, if the 22XT's voice coil (centered in the magnet) is about 6 inches behind the Pro-12 voice coil, then a 0.46 millisecond delay (0.5 feet/1100 feet per second) can be added to align the drivers. Note that the crossover frequency had no impact on the calculation, just the misalignment distance divided by the speed of sound.

TWEAK-QUALIZATION

The nicest thing about digital signal processors is that they can make a graphic equalizer virtually obsolete on main mixes. Of course, if you have guest mixers a sacrificial graphic equalizer is a good idea to keep their hands off your speaker processor that is tweaked-in to perfection. By using your ears, an RTA/ pink noise setup or a SMAART software system, you can flatten up the response to your needs or add that "special sauce" room curve that brings out the act's best show. Besides precision CD horn equalization, common parametric filter corrections are applied to mid-frequency hills or valleys and HF horn resonances. Also, adding Bessel or Butterworth response high and low pass filters at reasonable upper and lower driver frequency limits helps save your drivers from un-needed signal reproduction that is annoying at the least or destructive at the worst.

With digital signal processors, the sheer quantity of parameter choices is daunting for many of us. When in doubt, choose Linkwitz-Riley crossover point filters with 24 dB/octave slopes, just like most external analog crossovers implement. If you are transferring filter settings from one brand of processor to another, beware that the filter bandwidths are not always stated in the same terms. Some will use "Q" or Quality Factor in defining filter bandwidths, while others will use dB/octave. These representations all get to be brain mush when implementing low Q (wideband) filtering, so use your eyes and ears first until you are up to the task of understanding the mathematics of filtering.

5

THE MATING GAME, SPEAKER & AMPLIFIER MATCHING

April 2004

It always drives me nuts to walk into an install club and see two pairs of JBL SR4719 sub-woofers (each a 2 by 18-inch unit) powered by a single Crown Macro-Tech MA-5002VZ power amplifier. I know the amplifier can handle the load, and the speakers can handle the watts, but the match puts both pieces of gear in a non-optimum operating condition. Another club install rig I ran into has well-used, but well-maintained stage monitors of a music-store (MI) quality. Unfortunately, the chosen QSC PLX-1202 amplifiers barely get close to the wedge's RMS ratings.

In this chapter I want to get down to basics, because most of us still do not "get-it" when matching up sound system speakers and amplifiers. I hope to appeal to your common sense via a couple case studies. So let's go over rule of thumb on matching, and answer the SR4719/MA-5002VZ quandary.

PULL YOUR THUMB OUT ...

The rule of thumb on speaker driver matching is to size the amplifier maximum power rating in the range delineated by the RMS (continuous) rating to the program rating of the driver. Leaning on the RMS side is what I call the "cheapskate" side of the problem, and should only be a temporary solution until a bigger — or additional — amplifier helps with the load. The program power ratings for most speaker drivers are about twice the RMS ratings, and if all was perfect you would choose amplifiers precisely rated at the program power. Also remember that the driver impedance plays a big role, as the nominal rated impedance may better match up separately or together with the power amplifiers ratings, since they do not change in exact proportion to the load impedances.

In the example in the opening paragraph, the SR4719 sub-woofer has an RMS rating of 1200 watts and a program rating of 2400 watts at 4 ohms. The Macro-Tech amplifier offers 2000 watts at 4 ohms, and 2500 watts at 2 ohms in the dual channel mode. It seems obvious that one channel of the MA-5002VZ is a great match to a single SR4719, with everything operating near the program ratings. But the install cheapskates decided to avoid buying another MA-5002VZ for the second pair of SR4719s and now have loaded the MA-5002VZ at 2 ohms per channel. So now the amplifier offers 2500 watts per channel into two SR4719s, or only 1250 watts each — only 50 watts above the RMS rating. So now we have an underpowered sub-woofer with an amplifier taxed to the limit of its capabilities. Not a good formula for bass-heavy music like techno or hip-hop.

The other "nasty" I saw in this example was the usual single pair 12-gauge speaker cable running to the first SR4719 and then chained to the adjacent second SR4719. A "two-fer," or splitter Speakon adapter, at the back of each amplifiers output connector should have been used. Then each sub-woofer presents a 4-ohm load back to the amplifier, and the separate 12-gauge cables have minimal losses and a decent damping factor. Daisy chained 10-gauge speaker wires might have been an acceptable alternative.

Having a power amplifier slightly too large for the speakers's program power rating is not a crime, but it is something to watch out for in operations. Sub-woofers can be pushed into mechanical over-excursion if driven hard. Voice coil damage is possible, but less likely than over-excursion. The risk in under-powering speakers is that overdriving the amplifiers creates clipping or highly-compressed (limiting) scenarios in which the amplifiers deliver extra power and over-heat the voice-coils of the speaker drivers. This results from the fact that if a speaker cone runs out of linear excursion too long, while the voice coil receives maximum power input (heating), the voice coil has little thermo-dynamic means to cool itself because of the lack of much end-motion and sustained proximity to the heat sinking metals of the magnetic motor assembly. This heat build-up will eventually melt the weakest part — the voice coil conductor — thus opening the circuit.

CASE ONE

In case study one, we depart from sub-woofers and go topside. Suppose you have a pair of JBL SR4732 top-box speakers and you want to assess which amplifiers you could choose to power them. If you chose the passive, full-range mode on the speaker you have a 1200-watt RMS power rating. So amplifiers providing 1200 to 2400 watts per channel into 4 ohms will get the job done. If you go for bi-amping, the same applies to the low frequency split on the SR4732s. This is because the pair of JBL 2206H 12-inch LF drivers each handle 600 watts RMS in the recommended 100 Hz to 1.2 kHz frequency band. On the high frequency split, the answer requires a little intuition and knowledge of the MF and HF drivers. The SR4732 employs the JBL 2447J MF driver with a 16-ohm nominal impedance, and the JBL 2404H driver with an 8-ohm nominal impedance. Because the 2447 handles the 1.2 kHz to 6 kHz band and the 2404 "baby butt" handles the above 6 kHz band, the 2404's extra sensitivity is passively matched back to 16 ohms across the whole 1.2 kHz to 21 kHz split range with a 150 watt RMS rating.

Because power amplifiers normally do not offer a 16-ohm rating, one can conservatively state that one-half the amplifier's 8-ohm rating applies. So the upper split of the SR4732 requires 150 to 300 watts at 16 ohms, or about 300 to 600 watts at 8 ohms. With 16-ohm loads high and 4-ohm loads low, do not be surprised that the same amplifier can apply to both bi-amp loads.

If we start optimally sizing amps for the SR4732, typical good picks would be like the Crest Pro9200, Crown MA-5002VZ (a touch high) and the QSC PL-6.0II. These amps hit the optimum spot near the 2400-watt program power rating at 4 ohms for the LF split. For the HF split, amps like the Crest Pro7200, Crown MA-2402 and QSC PL230 would be among the candidates. Switching to cheapskate mode, the Crest CA18, Crown CE4000 or QSC RMX-4050HD would get you into adequate operational territory on the LF split. On the HF split, the Crest CA6, Crown XLS-602 or QSC RMX-1850HD would be on the good side of the 300 watts per channel at 8 ohms.

CASE TWO

Now lets flip things upside down and presume you have one QSC RMX-1450 amplifier to drive stage monitors with. If your wedges are going to be Yamaha Club Series V models, such as the SM12V, then the 250 watts RMS and 500 watts program at 8 ohms applies. The question turns into: How many wedges on each channel "should" you load on the RMX-1450? Taking into account the RMX-1450 ratings of 280 watts at 8 ohms, 450 watts at 4 ohms, and 700 watts at 2 ohms, yields part of the answer. With a single SM12V per channel, the 280 watts just barely gets you into manageable safety. With two SM12Vs, each receives half of the 450 watts, or 225 watts per wedge. This 25-watt underage is what can be considered playing with fire, or the safety of the voice coils.

With four SM12V wedges, the 700 watts is divided four ways for 175 watts per wedge. This power rating may be passable for political speech rallies, but never close enough for rock 'n' roll. This is where the repair shop loves you more than you love them. To resolve this dilemma, consider upgrading to the QSC RMX-2450, which offers 500 watts at 8 ohms, 750 watts at 4 ohms and 1200 watts at 2 ohms. Or better yet, get more amplifiers to drive fewer speakers per channel, with the advantage of separate mixes for each monitor chain.

Upgrading to the QSC RMX-2450 in our example, consider that a single SM12V per channel hits the program power mark at 500 watts. With two SM12Vs per channel, the 750 watts divided by two provides a moderate 375 watts per wedge. Even with four wedges, the 1200 watts by 4 leaves an acceptable 300 watts per SM12V cabinet. The caveat at 2 ohms is that the RMX-2450 cannot sustain continuous clipping in that loading, and this also applies to many other brands with similar power ratings.

FINAL NOTES

I can hear the obvious rebuttal "we have run under-powered for years and not blown up much." Running on the under-powered edge is truly penny-wise and pound foolish. The few extra dollars you have to wait and earn before choosing a larger power amplifier are well worth it in the long run. As sound system operators we all fool ourselves with the thinking that the brand will protect us from failures. All speakers and amplifiers will fail eventually, you should do what you can to make it an age problem and not an abuse problem.

Also, I humbly apologize to those speaker and amplifier manufacturers that I failed to recognize in this chapter. It does not mean that you are less in my opinion, but by indiscriminately choosing my examples, the readers can substitute your products by focusing on the power ratings and impedances.

Another Holy Grail many sound persons chase after is the concept of an auxiliary send feeding sub-woofers (aux fed subs). Traditionally a tactic of big arena shows, and sometimes used as a performance special effect, aux fed subs still makes sense for us smaller show soundco's.

WHY? WHY NOT?

First, we need to grapple with the question of change. One of the reasons why we should not use aux fed subs is that simplicity is wonderful. We've got top-boxes and sub-woofers, and a crossover in-between — why change? For small, budget conscious sound providers the thought of just going conventional and hooking up normally is comforting. Another reason to avoid this tactic is that aux fed subs now require an extra electronic crossover/speaker processor just to block the high frequencies and the pass the lows. The last reason is that some sort of calibration procedure is required to bring the aux/sub gain back in line with the main mix. And if your subs are not co-located with your tops, getting the phasing to match up requires an audio measurement tool like SMAART or equivalent time-domain audio tools. Sometimes you can run a test tone at the crossover point and tune the low/mid delays by ear to get the maximum coupling.

The positives to going to aux fed subs mostly revolve around clarity. By picking and choosing the mixing console channel strips to route through the subwoofer auxiliary send, the subwoofers sound remarkably clean compared to a conventional top/sub mains split. The reason for this is that channel sources (mics) that do not have low frequency sources still collect bleed from other parts of the performance area and create additional low frequency mud. Another reason is that switched, or swept, high-pass filters are not the brickwalls we like to imagine them. Most of these filters are two or three pole (12 dB/octave or 18 dB/octave), but even with –3 dB around 100 Hz, plenty of 70, 80 and 90 Hz signal gets by the filters. When many channels are summed in groups and on to the mains all that partially attenuated rumble is still audible. And don't get me started on stage floors or platforms with subwoofer resonances.

HOW TO

Now that you have had fair warning, here are the steps to get into aux fed subs. First, another speaker processor is going to be required. If you already have a stereo processor, you may be able to get by with a mono mix by feeding the main mix on one side and getting your top-box feeds from that. Then take your chosen aux send channel and feed the other side for a sub-woofer only mix. When deciding on another speaker processor for the aux fed subs, you may be able to take a chance on a more bargain model/brand since the low frequencies tend to be less critical then the highs on audio fidelity. Just make sure to get reasonable control on the gains and phase (delay) of the outputs. See **Figure 6-1** for a typical aux fed subs setup.

When wiring to the power amps, the sub-woofer amplifiers now get re-wired from the mains crossover/processor to the new speaker processor. This usually means another drive-line used up in your snake, unless you have the crossover/processors located at front of house. For simplicity, I recommend keeping the processors co-located in the mains amp racks. Dial the processors initially like they are identical setups. In other words, if you have been using a top(mid)/sub crossover point of 100 Hz and a Linkwitz-Riley filter of 24 dB/oc-tave, initially stay with those settings until you need to adjust the aux processor gains and phase response (delay).

IN PRACTICE

When using aux fed subs, typically the selected channels needing sub-woofer feeds have a post-EQ, post-fader assigned auxiliary send control at the unity gain setting. Depending on your console, this may be a setting from half-up to three-quarters of the way to full travel. The reason for this is that all the sends will maintain the balance as set by the faders. Of course, this advice can be ignored under the "if it feels good, do it" mantra. Setting or calibrating the master aux send control or fader should follow your group and master gain structure practices. On non-VCA boards, if you typically back off 10 dB on the groups and run your masters at 0 dB, try a nominal –10 dB setting on the aux send master for starters. But just like setting sub gains on main crossovers/pro-cessors you may either goose that setting or back-off, depending on music style or sub-woofer locations.

Doing aux fed subs on VCA consoles is much more a walk in the park, as the channel VCAs get gain cues from the VCA groups. This means less of a math problem, and more of an autopilot function as the groups are fiddled with. On non-VCA consoles, you may have to periodically goose the aux master control feeding the subs if you are in the habit of bumping the masters or groups dur-ing the show. In this case, remember a cardinal rule of mixing: Making some-thing louder can also be done by making everything else softer. A rule I wish more of us would exercise more frequently.

The oft-asked question is, "What signal sources are prime candidates for aux fed subs?" The quick and easy answer is kick drums, floor toms, bass guitar, some keyboards, some seven-string guitars and your break music decks. Again, the "if it feels good, do it" mantra can be enacted, but keep in mind why you are weeding out channels in the first place. Aux fed subs may force some tech-niques that smaller consoles do not have the luxury of handling. One example is that my stereo playback deck now has to reside on the channel strips to gain access to the subwoofer aux send, and this eats up precious and limited perfor-mance channels. This may be resolved if you can swap an effects return up to the console's master section if possible. Most effects returns, such as common reverbs and delays, can be neglected in the sub-woofer feed. Of course, an exception applies when I want to select my favorite "Devil Voice" pitch shift effect. Darth always sounds better with sub-woofer support.

A last question is the crossover frequency of choice when choosing aux fed subs. The best choices tend to be in the 90–120 Hz region, and are mostly a trade-off on your top-box low-mid frequency capability and the signal types you will chop off their low frequencies. A good choice with 15-inch drivers in the top-box is 100 Hz, primarily because six string guitars have a fundamental low E-string frequency of 82 Hz. Losing the fundamental harmonic on the first couple notes on a guitar is usually not noticeable in a rock 'n' roll performance because the harmonics tend to fool your ears into hearing phantom fundamental notes. Pushing your crossover to 120 Hz and beyond starts getting into the zone where you and the listeners begin missing depth in vocals and some non-sub instruments.

CRYSTAL BALL

My crystal ball foretells of a future in which power amplifiers will have DSP-based preamps with digital remote control. These features will usher out the extra speaker processor requirement as the amplifier with be programmed with its own audio pass-band. If each amp has remote control via a computer at FOH, then it will be possible to optimize the gain and phase delays for the performance venue using the measurement tools.

Figure 6-1

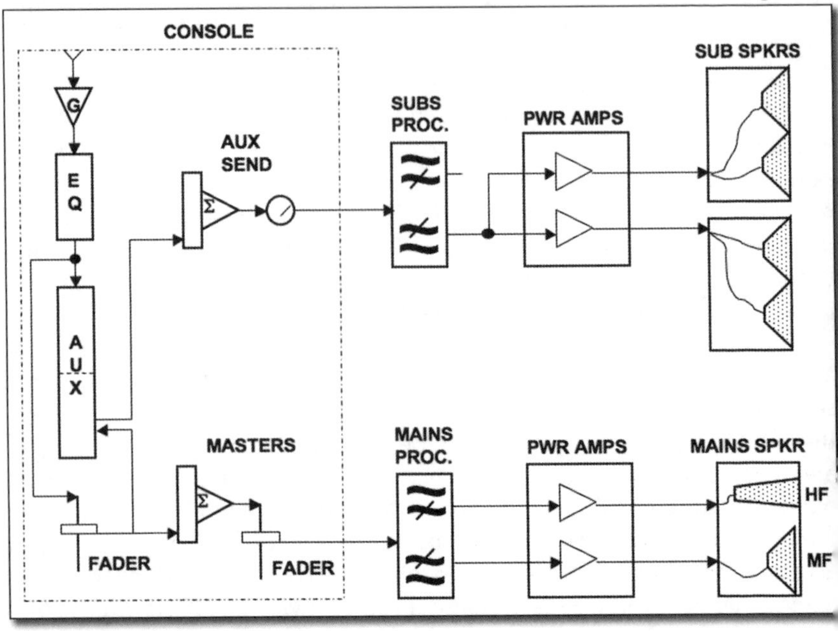

Frequency crossover networks have been around since the early days of movie theatre. Today we have active crossover networks in addition to legacy passive networks for splitting out bands of audio signal to frequency specific drivers (speakers). This chapter will go backwards in history by first discussing crossovers, then working through the basics of passive crossover networks for a two-way speaker system.

ACTIVE OR PASSIVE?

With the ever-dropping price of digital signal processing (DSP) speaker processors, the previous argument of going active if you could afford it goes by the wayside. Even with legacy active analog crossover units, the reasons to remain using passive crossover networks boil down to simplicity and the lack of need for a dedicated high frequency driver amplifier. As of this writing, several speaker manufacturers, both professional and non-professional, are offering low to moderate cost DSP-based speaker processors that perform the active crossover role, plus equalization and time alignment.

The principle of active crossovers is to pre-split the audio frequency bands while they still are line levels in order to avoid having to do the same while handling hundreds of watts headed to driver voice coils. This avoidance is driven by a few items, first amongst them is that the inductors and capacitors that make up passive crossover networks would absorb precious watts of power that should be pushed out to the audience instead of making heat. Another is the reduction of damping factor that impairs the driving amplifier for precisely controlling the motion of the driver's voice coils, thus blurring the fidelity of the speaker system. The last reason is that active crossovers can easily create very tight filters that could not be efficiently or cost-effectively implemented in a passive crossover network. Try implementing 100 Hz center 4-pole (24 dB) Linkwitz-Riley network with capacitors and inductors and you see my point.

So it is preferable you do the frequency dividing task at line levels between the console and the power amp racks, with either analog or digital-based active crossover networks. Of course, you have now doubled or tripled the number of amplifier channels you need to implement a bi- or tri-amped speaker system — not to mention the multi-pair cabling to the speaker via Neutrik Speakons or EP connectors, and the balanced patch cords from crossover to amplifiers, and the jack plate system in your amp racks to make everything plug, power and play. And don't forget the extra power distribution breakouts for the crossovers and extra amplifiers.

I made the difficult plunge from passive to active in my entire rig, but I took it in affordable chunks. The easiest, but never the least expensive, way was to get the amplifier acquisitions done first and make do with low cost analog crossovers until the DSP versions came along with the right configurations at the right prices. I have a BSS Mini-drive (FDS-336) for my mains speakers, and I am still using the industry renowned TDM 24CX4 quad two-way analog crossovers to create four bi-amp monitor mixes per unit. The version of the 24CX4 I have also has a simple but effective adjustable limiter circuit to help protect the amplifiers and drivers from excessive signal levels. The obvious next step is to retire the 24CX4s for a DSP unit that will include time alignment and preset equalization.

PASSIVATION

The reasons for sticking with the legacy passive crossover system also have to do with the fact that most speakers for music industry (MI) and pro-sumer customers already have the passive crossover network included. So why not save on cable and electronics and just leave it alone? But unless you build your own speakers already, you probably wonder what makes up a passive crossover network in speakers these days.

Whether you are building your own passive crossover network using purchased parts, or just want to know what technology is inside your own cabinets, let's explore the basics using a simple two-way, two-pole, passive crossover network. **Figure 7-1** shows schematically such a network with the drivers and cabinet jackplate interface. The two pole description means that two filtering components (capacitors and inductors) are used to separate frequencies with each "pole" meaning a -6 dB per octave slope contribution in the crossover network.

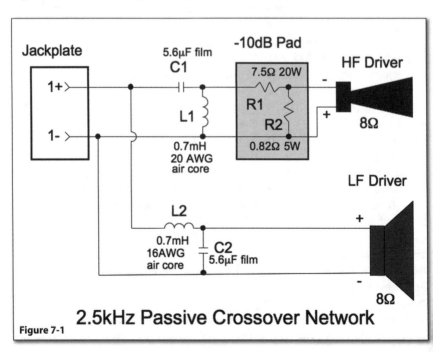

2.5kHz Passive Crossover Network

Figure 7-1

In two-pole designs, the woofer (low frequency) circuit gets a series inductor and a shunt capacitor to reduce high frequencies and pass low frequencies. Likewise, the tweeter (high frequency) circuit gets a series capacitor and a shunt inductor to reduce low frequencies and pass high frequencies. Most two-way speaker cabinets will employ a passive crossover like this. Where things get artistic is how the manufacturer's designers choose the component values to set the crossover point and how much dip the driver networks intersect at. A classic example would be where a 12-inch woofer and a 1-inch throat tweeter intersect at a crossover frequency of 2.5 kHz with the crossover network chosen to create a -3 dB (half power) loss for each driver at the 2.5 kHz frequency. At that exact frequency, each driver shares half the reproduction burden of creating the sound wave.

MATCHING MAGIC

You may have noted the flip in the tweeter polarity in the **Figure 7-1** example. The reason for this flip is that at the crossover frequency both parts of the network introduce a phase error and combine nicely when one driver polarity is flipped. With active crossover systems, this flipping is not needed, and can be undone if there is active/passive switch on the cabinet jackplate. But this gets proprietary with manufacturers in the sensitivity matching part of the passive crossover network. I am not going to reveal anything related to a specific manufacturer, but the figure does show a basic two power-resistor "L-pad" configuration as an example of what I would do for home-brew design.

You may have noticed differing woofer and tweeter sensitivity ratings in speaker specifications, with the high frequency driver 6 to 10 dB more efficient. So when using a single amplifier, passive networks must pad down about the same number of dBs on the tweeter network to match up these sensitivities. But no one said that you had to use power resistors exclusively for this matching. In fact, most designs for matching employ proprietary mixtures of capacitors, inductors, resistors and low voltage light bulbs.

And then there is the raging controversy of what tweeter protection designs are best. Philosophies range from absolute circuit breakers to softer multi-step protection, using things like light bulbs, poly-switches, relays and circuit breakers with standard components. While I fall upon the KISS principle to just breaker the tweeter, manufacturers will expend plenty of marketing dollars to let you know how much torture their products will endure with their crossover network driver protection schemes.

If all of this inspires you to try your hand at brewing your own passive network, check out a catalog supplier called Parts Express (partsexpress.com). They specialize in drivers and network components for the do-it-yourself enthusiast. They have multiple pages of tips and techniques to help you select the best network for your drivers. My best advice is not to skimp on inductor wire size and go with film capacitors, plus do not go the "audiophile" route with esoteric components. After all, back in the 1970s, all we had for crossovers were oil-can starter-capacitors normally used in window air conditioners.

How some people position their loudspeakers has been a pet peeve of mine for quite a while. For outdoor situations and large, wide rooms I can go with the flow of whatever provided system I mix on, but there are many occasions, especially in narrow or "tight" rooms, when I come away shaking my head about speaker positioning. This chapter will detail some basic things to consider about speaker positioning.

FLAT FRONT VERSUS COCKED

When the performance area is a narrow room, or a partially segregated performance area where the loudness will be confined, the typical speaker stacks on the side of stages should be slightly tilted or "cocked" inward. Of course this may be a bit fruitless if you have been provided music-store-grade top-boxes with 90 by 40 degree horns. In that case, just cock them 15 degrees and prepare for wall splatter reverberation.

As you might be able to infer, I am a big fan of medium to long throw cabinets with at least 60 by 40 degree horn patterns for the high frequencies. With the exception of small clubs I am not fond of wide pattern horns. **Figure 8-1** shows what I am referring to. Also, many times there are venues where I want areas with no coverage, as some patrons may want quieter side areas so they can converse during the show.

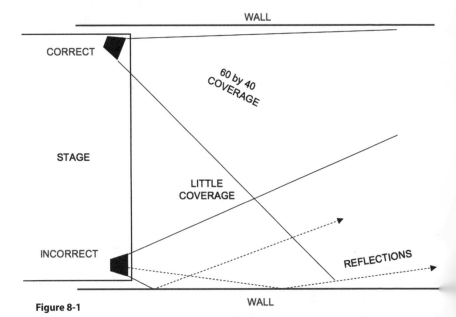

Figure 8-1

MORE SPL AND MORE SPEAKERS

With the exception of subwoofers, the call for more loudness does not necessarily correlate to more speakers. Usually, more efficient speakers are the best practice for medium-sized venues. By adding more top-box speakers to each side of a stage the differences of sound wave arrival times from each of the co-located cabinets starts blurring the fidelity of the mid and high frequencies, if not causing outright comb filtering.

The best way to solve the "more loudness" problem is to select speakers optimized for efficient medium- to long-throw cabinets. There are plenty of 98 and 99 dB SPL at 1-watt, 1-meter top-boxes out there, but doubling up on these boxes yields only a 3 dB improvement — assuming the phasing issues are solved for minimal combing and blurring. So why not choose medium-throw cabinets like the popular EAW KF650z (101 dB SPL) or the EV QRx212 (102 dB SPL) and get that extra loudness for nearly no extra cost? And if you really want to double the loudness (+10 dB), try the Community Solutions SLS960 or SLS980 long-throw top-boxes with 108 dB SPL sensitivity.

By staying with single tops per side, not only do you avoid hauling a second set of tops around, but there are less speaker cables, amplifiers and less power to distribute in those all-too-common difficult venues that have little juice to spare. And hauling less racks and stacks is just fine to those paying the fuel costs, and make that double for the crew humping the gear.

Now, there are common situations where multiple tops are appropriate. That gets back to wide audience areas where you can splay two or three medium-throw cabinets. If I am provided a rig with double tops and the room is not wide, I will either splay in the inside tops **Figure 8-2**, or just disconnect the outside tops and leave them just for aesthetics. Of course, there is always the image problem perpetrated by bands or venue management that must have "a lot" of speakers on each side for the perception of loudness — but no one said you had to cable them all up.

If you truly need more loudness than a single pair of speakers provides, then consider "inverting" a second pair of tops on to the first pair **Figure 8-3**. This inversion works best if the cabinets can be secured, and if the cabinets are two-way (mids/highs). Three-way cabinets (lows/mids/highs) probably will not work if the mids are placed between the high horn and the low drivers, because the distance between the mids becomes more than a half wavelength and the infamous phasing issues arise again for blurring/combing. When you invert top cabinets, usually the high frequency horns couple together and act as a single louder horn.

SUBWOOFER ARRAYS

The classic quandary with subwoofer cabinets is whether you should stack 'em up on the sides of the stage or center them beneath the stage for a horizontal array effect. The rationale for side stacking is mostly practicality, as the

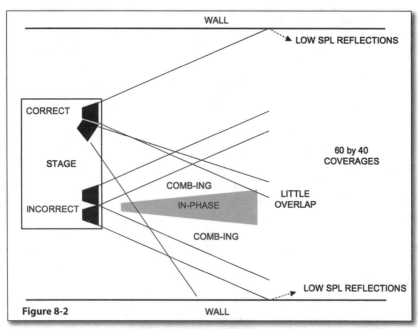

Figure 8-2

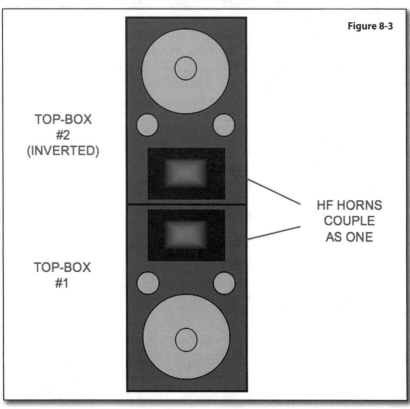

Figure 8-3

top-boxes need a height boost — after all, if you don't fly them, you've still got to get them over the audiences' heads. With 10-foot wavelengths, close coupling multiple subwoofer cabinets does a fine job, but two piles of subs stage-right and stage-left can still create comb filtering effect as they sum together in the center of the audience. That is true if you are within a couple feet of the center-line between the sub stacks.

Laying subwoofer cabinets end-to-end can create that awesome looking horizontal line array, with the benefit of more evenly spaced loudness across the stage front. However, there are always exceptions. For example, I know a club where the main bar is an island 75 feet out from stage-center. With the subs arrayed on the floor their rumble pretty much pisses off the half-deaf bar staff at anything beyond Lawrence Welk Champagne Music levels. At least the sound company provider had enough sense to fly the medium-throw tops and aim away from the bar.

So if two strong lobes of low frequency support are the best, then side stack the subs. If a broad uniform forward lobe is best, then array them across the front. And don't forget the free +3 dB hard floor bounce efficiency you get when the subwoofers are on the ground and not on high stage decking — if you stack the subs vertically the higher stacks get even less floor bounce advantage.

One of the things that still aggravates me are badly formed questions about loudness — "How many amplifier watts-per-person are needed for a rock 'n' roll show?" And the oft-mentioned "How many speakers do I need for this size room?" While the innocent are allowed to ask such questions, I find it disconcerting to hear experienced sound persons asking the very same questions. In this chapter I shall attempt to cover the basics on estimating the sound pressure level (SPL) provided, and then reverse the basics to show you how to estimate the amount of gear need to achieve a desired SPL.

Before we start, let me state that there is virtually no such thing as uniform SPL, short of handing everyone personal MP3 players with the volume controls fixed. Once you decide on speaker location, common sense tells you it will be louder near the speakers and softer further away. You can, however, specify the range of SPL. That will drive the capabilities of the speakers, and maybe even require more speakers dispersed into the audience if the range is tight. But in most small venue situations speakers are located above or to the side of the performance stage, and no capability or accommodation will be made for additional speaker locations.

SPEAKER COVERAGE

It's not all bad news if you are limited to speaker stacks at the sides of the stage. As many of you know, dispersed speaker positions require an electronic delay of the mix audio to get the stage wash and side-stage speakers to align with speakers placed further away. Given your favorite speed of sound number for temperature and humidity conditions (about 1100 feet/second), just back off a delay of about 0.9 milli-seconds per foot of distance from the stage front for each dispersed speaker. Of course, the downside is the complexity of having to route delayed signals to the speakers and determine the HF horn coverage to the zones desired.

In smaller and straightforward room shapes, the side-stage speaker stacks can do all the work. For smaller rooms, the traditional speaker on a stick (tripod) with the 90 by 40 degree horn is the standard. In bigger and louder applications, medium-throw speaker enclosures use 60 by 40 degree horns enclosed in trapezoidal cabinets for throw of 100 feet or a bit more. In wider rooms, two or three medium-throw cabinets are splayed together to widen out the coverage and still keep the SPLs high. As in the dispersed speaker setup, the idea is to have one speaker focused to each audience section, with other speakers not covering the section by virtue of being out of the rated horn dispersion of the cabinets. This way the dominant SPL source does not have significant competition at mid and high frequencies.

If you did not now it already, most subwoofer cabinets offer very little dispersion pattern control below 100 Hz. With this known, most subwoofer cabinets are close to the stage — either onstage, below the stage or side-stage as space is available and the audience locations are factored in. With wavelengths at 10 feet or longer, massing identical cabinets together forms a virtual single cabinet with the same sensitivity but with the input power effectively summed together.

INVERSE SQUARE LAW

To get at the problem of delivering SPL into an area, I have created **Figure 9-1** to show how SPL diminishes with distance from the "point" source. Since a speaker or small cluster of speakers aimed together represent a single point source when looked at from afar, the resulting acoustic wave acts spherically and diminishes its intensity as an inverse square with distance. With the speakers as a transducer, dB electrical watts in correlates to dB SPL out via the sensitivity rating of the speakers. In **Figure 9-1**, a 102 dB SPL per watt per meter sensitivity cabinet with 100 watt electrical input (20 dB watts) equates to 122 dB SPL at 1 meter.

From the figure, each doubling of distance away from the speaker drops the SPL by 6 dB. So 122 dB SPL at 1 meter equates to 98 dB SPL at 16 meters, about 52 feet away from the speaker. The only way to avoid this rule is to get into line array speaker systems for cylindrical dispersion and 3 dB per distance double losses. But even with line arrays, once you get below the critical low frequency, the low frequencies begin to drop off faster just like in the spherical dispersion model. Then you have a real need to add delay low frequency cabinets to keep up with the mids and highs screaming along in the cylindrical dispersion rate.

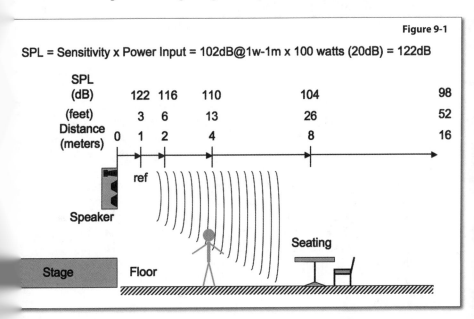

Figure 9-1

SPL = Sensitivity x Power Input = 102dB@1w-1m x 100 watts (20dB) = 122dB

RUNNING THE NUMBERS

The old rule of thumb for rock concert SPLs in audience areas is to keep things in the 90 to 120 dB SPL range of loudness. With a normal conversation at 70 dB SPL and each 10 dB representing a doubling of perceived loudness, you can see that at concert levels, conversations can only be carried on by shouting into each other's ears. We can assume that most concertgoers will not get within a couple meters of the speaker stacks, so the first 6 to 12 dB of SPL loss from the speakers will not normally be hitting ear drums. So with medium- to long-throw speaker cabinets, it is likely that the maximum SPL capability may exceed 135 to 140 dB SPL at the 1 meter reference. Those levels are needed when the SPL at the audience seating area is 120 dB.

Looking at **Figure 9-1** and adding 10 dB SPL means that 132 dB SPL emanates from the speakers (1000 watts input), and at 4 meters the 120 dB SPL requirement is met. Doing the math with drops of 6 dB at distance doubles tells me that 92 dB SPL is still maintained at 64 meters (208 feet) away from the stacks.

Looking at the problem differently, suppose you wanted at least 90 dB at 100 feet away (about 32 meters). Then in the **Figure 9-1** configuration, you need 96 dB at 16 meters, 102 dB at 8 meters, 108 dB at 4 meters, 114 dB at 2 meters and 120 dB at the reference 1 meter distance. But let's say you have 15-inch+1.4-inch top-box speakers that only deliver 98 dB SPL at 1 watt and 1 meter. Then the 120 dB minus 98 dB sensitivity formula says you need 22 dB watts program into the cabinet from the amplifiers. Then you must take the 22 dB, divide by 10 (power is in 10-log dB scaling) and do the base-ten anti-log computation on 2.2 for 158 watts program power. Then if 90 dB at 100 feet suddenly becomes 100 dB, then you need 32 dB watts or 1580 watts of voice coil melting power. This is where two speakers sharing the 1580 watts (790 watts per cabinet) on each side of the stage works; or better yet, coupled together to minimize phasing.

FINAL THOUGHTS

While people do absorb acoustic power, they do not rob power (loudness) from adjacent areas. Thus the watts per person theory should be thrown out. And the "how many speakers" computation also falls away, as different speakers have different sensitivities/efficiencies and large quantities of the same speaker rarely couple together well, to place increased SPL into any one location or locations.

What does matter is the intensity of the emitter (speaker with amplifier) that is pointed in your direction, and your distance away from that emitter. Things can get complex in big venues with many emitters and coverage areas that can overlap. That is why the big shows do have system engineers, and they are in charge of the math and physics to deliver the correct loudness in the right areas, with the best sound fidelity.

POWER DISTRIBUTION

Power distribution topics have been very under-reported in the world of live sound — partly because of ignorance and partly due to the concern of the dangerous consequences if done incorrectly. Power distribution cannot be taken casually. Nonetheless, it is still a very important aspect of live sound operation. I recommend obtaining an electrician's license if you plan on working regional and national gigs. It'd be a good credential, not to mention a good safety tool.

I know of nothing more important to providing a good, reliable sound system operation than having a great power distribution system. But before we dive into the ins and outs of cam-loks, genny connections, star grounding and load sharing, let's review the basics to bring everyone up to a level of understanding.

Power distribution is not just a little rack mount unit that splits one plug into several receptacles, although you would never know that based on the sparse information about the subject. Most of this is because a little knowledge in the hands of the inexperienced makes those individuals very dangerous, and their reference materials could be subjected to lawsuits (read: lawyer bait). So, at the risk of baiting the legal community, here comes the information.

Safety is first, of course. Besides knowing how to implement a power distribution system, you should know that the result of doing it wrong opens up the possibility of shocks, unintentional welding, fires and shrapnel flying about. If you have the slightest thought that what you are about to touch may shock you, make sure you touch it with dry skin and with the minimal amount skin surface area. I recommend a single knuckle, in case the current stimulates the closure of your hand muscles (think palm on mic windscreen). Your dry skin may have thousands to tens of thousands of ohms of resistance, and is your primary protection against shocking currents. Once inside your body, the resistance drops to a few ohms. A GFCI (Ground Fault Circuit Interrupter) trips at 6 milli-amperes of current, which is still enough milli-amperes to give you a tingle. If your muscles get vigorously stimulated from a shock, get yourself checked out by a medical professional. For a full rundown shocking current levels and their physiological effects — from tingles to muscles ripping off bones — check out the chart at the beginning of the book.

If you plan to start constructing power distribution gear like portable power distribution units (PPDUs), backline stringer AC cables and feeder extensions, getting a copy of the latest National Electric Code (NEC) handbook is a must. The book details the terminology, practices, components and location-specific regulations needed for the safe and legal distribution of electricity. Get yourself a personal copy (www.nfpacatalog.com), or borrow one from the local library until you can afford one.

While not widely enforced, all work on power distribution systems should be done by licensed electricians or licensed electrical engineers. The coursework to become an electrician is taught at vocational-technical schools as a two-year degree, and would make an excellent credential to supplement your sound system experiences. While the license is only valid in the state you are tested at, the experience and the credential will help on tour when dealing with other building electricians. If you do not pursue licensure, at least read the NEC handbook a couple of times, and have a licensed electrician/engineer double-check all your work.

WATTS UP DOC?

As an acknowledgement for James Watt's work on steam engines and defining the amount of work (power) on them, physicists have defined a standard unit of power as a "Watt" instead of the previous usage of horsepower. For electricity and power distribution, knowing "Watt's Law" is as crucial as knowing Ohm's law. Watt's law is simply:

Power (Watts) = Voltage (Volts) × Current (Amperes)

Here in the good ol' U.S.A., our AC power for sound systems is mostly 120 Volts alternating current (VAC). Given that standard residential and commercial receptacles use 15 or 20 amperes, Watt's Law would say that the circuits that feed the receptacles would then be either 1800 or 2400 watts in power availability.

Watt's law does apply for many items, and the challenge is to distribute the various pieces of sound system gear so they do not add up beyond circuit power availability (capacity) or a potential "overload" may occur — resulting in a circuit breaker tripping or fuse blowing. There are obvious exceptions in sound system work, primarily in sizing audio power amplifiers to power distribution circuits. Because music reinforcement is not continuous in loading like DC or AC power circuits, audio power delivered to speakers does not directly transfer to continuous power supply consumption. For example, average music power delivered to the speakers typically is one-eighth to one-third the rating of the amplifier. Then you need to factor in the extra power lost in the form of heat in the amplifier circuits. That is why high-efficiency 4000- to 6000-watt rated power amplifiers can be successfully supplied from one 2400-watt circuit. Consult the amplifier manufacturer's specifications on power draw based on your nominal speaker loadings and intensity of operation.

THOSE PRETTY COLORED WIRES

To understand how power is supplied from the electrical utility down to your gear, we need to draw a picture. **Figure 10-1** shows a simple commercial or residential single-phase electrical service. While denoted as single-phase, it actually is two "hot" conductors out of phase with each other, sharing a common return wire that is called a neutral conductor. The electric utilities' transformer converts the many kilovolt level power line energy at the primary winding to a secondary winding of 240 volts with a center-tap point for the neutral feeder connection, which then creates two out-of-phase hot feeders. These beefy feeder wires exit the transformer vault or pole and run to the venue's meter socket, which becomes the end of the electric utility's responsibility for power provision.

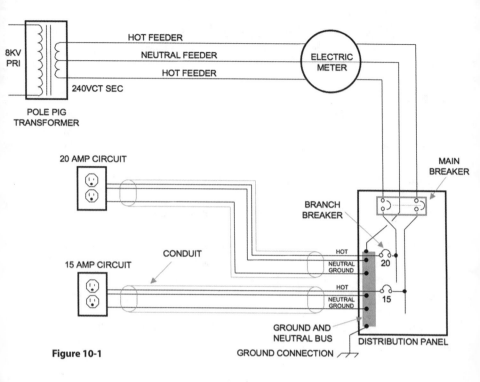

Figure 10-1

From the meter socket the feeders continue into the building location until they reach the main power distribution panel. Larger residential and commercial locations may have sub-panels to further group circuits by area, and have additional feeder wires between the panels. Besides the single-point entrance for electrical service, the main distribution panel serves as the point where the neutral feeder wire gets attached to a safety ground conductor wire that makes a good electrical contact with the soil surrounding the building. This neutral-ground connection is only supposed to be done at one point in the electrical distribution system, and is made in the main panel neutral bus bar. The bus bar is usually a rectangular bar of steel with many holes that serve as conductor inlets. These inlets have screws to fasten each neutral and ground conductor to the bar. The hot feeder wires also connect to hot wire bus bars that are connected to main circuit breakers or fuses to interrupt the circuit when very large overloads or short circuits occur. From these safety circuit interrupters, the hot-wire current paths connect to branch circuit main busses where they will be tapped by breakers or fuses serving smaller branch circuits.

The branch circuit wires that transfer electrical service to parts of the venue are denoted as hot, neutral and ground. Typically, the hot wire jackets will be black (phase A), red (phase B), blue (phase C), or violet (aux hot designator). The neutral wire jacket is white or gray, and the ground wire jacket will be green or jacket-less. In residential situations, branch circuit wiring may be covered with an overall plastic jacket (sometimes called Romex) that is white, yellow or gray in color. With some exceptions, commercial wiring is carried in metal or plastic piping called conduits, which lead to the outlet boxes that contain the final connections for light fixtures or receptacles

The wire sizing, or gauge, helps determine the maximum amount of current (or ampacity) the circuit can handle. Typically, 15-amp residential service circuits use 14-gauge wires. The lowest commercial service circuits are 20-amp rated and require 12-gauge wiring. At 30 amps, 10-gauge is used. For 50-to-60 -amp sub-panels and portable distro panels, 6-gauge wiring is required. Larger wire sizes are used if longer runs are required. The important thing to note is that the wiring, receptacles and any other items in the circuit must be rated as good as the breaker trip rating. Too many times I have heard some maintenance dude say, "I replaced the 20-amp breakers with 30-amp breakers, so I don't have to run to the panel as much when bands plug in their gear". The potential result is very warm wiring, smoking receptacles and a fire hazard.

IN CONCLUSION

It is my fondest wish to get everyone past the "dangerous stage" in working with high current power distribution, and practice safe connections per the NEC handbook recommendations. For me, demonstrating competence in sound system power distribution is the biggest factor in discriminating the amateurs from the professionals.

For many sound system operators, a common problem scenario exists: You are happily mixing away in mid-show and a circuit breaker overloads, causing an unscheduled set-break and leaving you with the quandary of how it happened in the first place. In this chapter, I will lead you through the appropriate loading of AC circuits and teach you to understand the causes of overloads. The best way to show is by example, and I'll use one that is vivid in my mind after many summers of civic outdoor street dances.

STREET DANCE REALITY

Your civic volunteer sponsored street dance gigs are a classic "production hell" situation. On one hand, these events are open air and with hundreds to thousands of attendees, requiring every reserve ounce of audio power and stage lighting to make it a memorable event. On the other hand, most electrical provisions barely border on adequate — which in this case means good enough for two accordionists and a tuba player.

Consider a real life small town street dance where I had one 50 A, 120/240 VAC receptacle (NEMA 14-50R) to do the whole bandstand — FOH power, instrument backline, main and monitor power amps and a modest amount of stage lighting. The standard contract clearly read "100 A, 120/240 VAC service within 100 feet of the stage," but the local officials handed me the well-worn line, "It always works for the other bands." Being resourceful, I looked up the power pole where the receptacle medium voltage (120/240 VAC) came from — and then I expressed my slight displeasure with a heartfelt "Drats," as I spotted nothing beefier than a 6-gauge feeder wire heading down to the breaker box feeding my receptacle. My hope of 2-gauge 100 A service to tap directly onto withered away.

LOAD SHARING — BY THE NUMBERS

As I pulled out my wimpy 50 A club portable power distribution panel (distro pack) and a 25-foot NEMA 14-50 feeder extension (6 gauge, 4 conductor); I began to ponder "50 amperes per hot-wire and three 20-amp breakers per hot to pull it through." With six 20 A breakers each feeding duplex Edison receptacles (NEMA 5-20R), the task is also bit more complex as I want to balance feeder currents to minimize neutral feeder hum and buzz, and just plain not have to worry about distro noises the whole gig. On the bright side I have a separate ground wire (from the bonded neutral bus in the main panel) and every branch circuit in my distro pack uses it. **Figure 11-1** depicts some of the thinking.

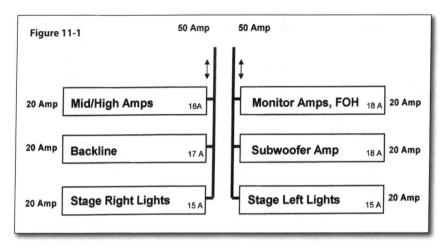

Figure 11-1

50 Amp	50 Amp

20 Amp	**Mid/High Amps** 18A		**Monitor Amps, FOH** 18 A	20 Amp
20 Amp	**Backline** 17 A		**Subwoofer Amp** 18 A	20 Amp
20 Amp	**Stage Right Lights** 15 A		**Stage Left Lights** 15 A	20 Amp

The first items I connected up were my main mid/high power amps and four mixes of monitor amps (driving four 12-inch+1-inch JBL passive wedges). Since my BSS minidrive, QSC PLX-3402 (bi-amp mids) and QSC PL218 (bi-amp highs) for the main speakers all ran through a Furman PM-PRO for protection, I had a LED bargraph visual for current monitoring from the FOH position. All this went on one 20-amp breaker, with the monitor mix amps (two QSC PLX-3402s) on the opposite hot feeder. My rationale was that both main mid-highs and monitors got very similar signal inputs, so placing them on opposite feeder hots would balance out the neutral current fairly well. Another fact was that each PLX-3402 is spec'd to draw typically 7.6 amps with 8-ohm loading per channel, and 11.6 amps with 4-ohm loading per channel. The PL218 was driving 8 ohms per channel, and typically drew 6 amps. So both branch circuits were under the load limits of 20 amperes. And because the monitor amp circuit was well under the limit I ran my FOH power from it, since it added only a couple more amperes.

The next two balanced circuits were the subwoofer amp circuit and the backline circuit. This will always be somewhat of a trial by fire, in that you may or may not know what the act's total backline power consumption may be. Yes, you can check instrument amplifier fuse or breaker ratings, but they are likely to 30–100 % higher than the nominal power consumption. Fortunately, I knew the whole backline was mostly combo amps, and would not challenge a 20 amp breaker, though they will occasionally trip a 15 amp residential breaker. The subwoofer amp I had for this gig (a Lab.gruppen fp6400) preferred the solitary use of a 30 amp, 120 VAC circuit to give the subwoofers everything it has. Fortunately, with a little backing off the kick and bass guitar low-end, it had proven to run reliably on a 20 amp breaker.

These are the kind of calculating things you do you to ensure a smooth show without outward display of stress to the performers and patrons.

Now, I beg your forbearance to briefly dive into stage light karma, even though this book is dedicated to live sound. Unfortunately, smaller productions do press the sound crew into running the lights when budgets are tight.

The last two circuits I reserved for stage lighting. Remember, I had 50 ampere main breakers at the power pole panel and my distro pack, and had allocated the better part of 40 amperes in the previous four breakers. In my rig I had 16 fixtures (four light trees, with four cans per tree), with ratings between 250 and 300 watts per fixture.

My stage light philosophy bends towards setting scenes, as opposed to the more traditional "flash 'em all" shows I see today. Fixture gel selection is red, blue and amber hues, with me using single colors or color pairs for variety and mood setting. I also make sure that Rosco 02 gel (bastard amber) gets liberally applied since it is a flesh tone that allows the audience to view the expressiveness of the performers.

With that said, all this meant that no more than two to three fixtures of any four-can light tree could be on at any one time for my street dance gig. Also, by chaining each side of the stage's front and back tree dimmer packs together on one circuit, each light circuit tended to have the same color washes on at the same time, thus further reducing any dimmer buzz and imbalanced neutral currents.

GOING FOR A TRIP

When loading up circuits, it helps to understand how a circuit breaker protects the circuit elements (wiring, receptacles, plugs and cables). Most common breakers use a thermal method determining when to trip. A small amount of resistance in the breaker creates a modest power loss that is converted into heat. The contact element that opens the circuit in a fault situation does so based on a buildup of temperature upon itself. Since the heat generated is proportional to the amount of resistance on the line, more resistance means more heat. Under normal circuit loading, surrounding elements wick away heat (heatsinking) to prevent too high a temperature on the contact element. What results is a curve of circuit current versus time in seconds before the breaker trips. On a typical 20-amp Square D Q0 breaker, 25 amperes of continuous loading would take about 2 minutes to trip, 40 amperes about 20 seconds, and 100 amperes about 2 seconds.

Given that you can draw more power than a circuit is rated at for a short period of time without tripping it, and that modern music reproduction loading tends to average 50 to 80 % of its peak current draw (we are talking about the AC side of power amplifiers here), most overloads tend to be gross infractions of load current estimation. Stage lights are the classic exception, in that "on" is full current draw, and they tend to have nasty turn-on current spikes when "bumped." These flashes or bumps tend to have about double to triple the spike current draw of their constant on current. Thus, the worst-case situation would be all fixtures pulsing on and off simultaneously; and doing this for many seconds could cause a breaker trip if the on current is close to the breaker rating.

CORDS AT YOUR SERVICE

One of my other pet peeves is the neglect for choosing the proper AC service cords (extension cords) for use in performances. The first faux pas I see is the substitution of orange or blue plastic extension cords for the desired black rubberized plastic service cords. The rationale for black is not only the better quality of material, but the psychological fact that black cabling hides well on stages and bandstands, so the audience is focused on the more colorfully illuminated performers and their instruments. Friends don't let friends use "ordangey" cords at shows.

Another blunder I see with service cords is the wire size or gauge used. Because gig to gig they may be used to the limits of breaker capacity, no smaller than 12-gauge cabling should be used, and 10-gauge is recommended if you need long runs at 20 amperes. I buy my AC cable in 250-foot spools and attach premium quality nylon plugs and receptacles to make up bunches of 100-, 50-, 25- and 12.5-foot length service cords. Another requirement that applies to theatrical stages, and should be applied to music performances, is the use of "extra hard service" (Type S) jacketing instead of "junior hard service" (Type SJ). The National Electric Code (NEC) requires all theatrical stage service cords of 20 feet or longer to be of the extra hard service variety. Look for designations like 12-3 SOW instead of 12-3 SJOW when cable shopping. The OW suffix stands for oil-proof and weather-proof. The rationale for extra hard service is because we know those street dance gigs will have vehicles, roadcases and people trampling all over them until they are packed away for the next gig.

In response to a column on power, a reader once asked me: "How do you hook up your system to a generator correctly so you don't have to worry about safety issues like damaged equipment or a ruined show? Most professionals have unique methods and they all look extremely dangerous."

Tying in to a portable generator is similar to tying in to house power, but there are extra tasks for both the sound and lighting production providers (load side) and the generating equipment operator (source side).

I equate making power distribution connections similar to doing medical surgery: if done correctly, by experienced professionals, everything will go well with no issues. But the incorrect way has very serious consequences, including extensive damage to gear, serious injuries and possible loss of life due to electrocution. That is why I can only endorse licensed electricians to do this work, and especially electricians that routinely do portable generator tie-ins. These specialists are typically concert venue electricians, tour electricians and senior regional soundco personnel holding an electrician's license.

Now that I have done my obligatory scare paragraph, if you plan on making electrical connections, please commit to memory most of the contents of the National Electric Code (NEC). The NEC books are updated every few years and available through the National Fire Protection Association (www.nfpacatalog. com). These rules and regulations guide how generator connections are to be made, and also give essential information on grounding, wiring, interconnects and specifics for indoor (theatrical) and outdoor (carnival) live music performances.

CAM-LOKS

Most generator tie-ins will be made to your power distribution rack, and done via individual wires, called feeders, that are not jacketed together but kept bundled by "banding" every few feet with tape or other means. Most feeder wires are black rubberized and weatherproof insulated, and vary in size from 2 gauge to 4/0 (four-aught) gauge. Most of these feeders are terminated with high-current-capable, weatherproof, twist-lock connectors called Cam-Loks, and retain a color on the connector rubber jacket (boot) that corresponds to the wire's function. For example, green boots indicate safety grounds, white boots for neutral wires and black, red and blue boots for hot wires. For sound and light production, all the hot wires are set to 120 volts AC with respect to the neutrals and safety grounds. Although Cam-Loks come in three sizes, the middle size (type 1016) is the standard type used most often.

Of the generators provided for performance use almost all will have a single or three phase switch for load selection. Smaller production companies will tend towards single-phase use at 240 volts AC (120-0-120), as most of their power distribution is geared for 100 amperes or less, and any house power gigs are also in single-phase. Three-phase systems tend to be run by larger regional or national soundco's, as they need the extra power transfer that this system offers. In single-phase systems, a four-wire loom of feeders is usually employed (green, white, black and red). With three-phase systems the third hot wire is added (blue) and 208 volts AC can be measured between the hots (120 VAC to neutral). Larger ampacity power distribution systems will double up on neutral and safety ground feeders (7-wire) as the NEC requires neutral and ground ampacity to be double any one hot feeder in three-phase systems.

Some generators may not use Cam-Lok connectors for feeder termination, so you may wish to have a spare 10- to 25-foot set of feeder "tails" that have bared conductor ends for buss strip connections on the generator's output panel. In this case, have your electrician make these connections and ensure no live voltage exists on the busses while making the connections. Even with Cam-Loks, meter the voltages coming out of the operational generator before connecting your distro rack to the feeders. Also make sure to have spare ground and neutral Cam-Lok "turn-arounds" (female to female adaptors) as some generator (genny) carts may have male Cam-Lok receptacles on the output panel. Do not rely on the genny rental house to have these handy adaptors. See **Figure 12-1** for a typical single-phase genny cart to distro rack setup.

Figure 12-1

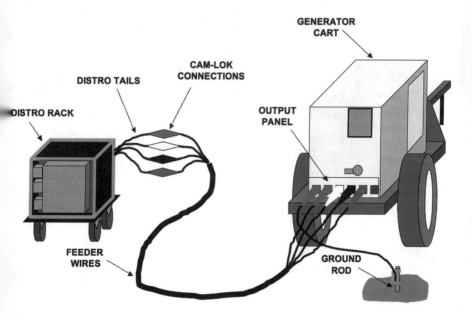

CHECKLIST

When employing a portable generator for a gig, the soundco and genny rental house should coordinate via the phone before the event to make sure everything will be the right configuration before the generator is set up on the sight. Ideally, the soundco personnel should get a brief training on generator operation before the rental company personnel depart for the next task. Sometimes all you can hope for is that the generator is sitting at idle, nicely earth-grounded, fully fueled and ready to tie-in. The following is a typical checklist that you and the genny operator should run through:

1. Check fuel, oil and coolant levels for proper levels. This is usually an operator task.
2. Select single or three-phase operation, 60 Hz operation, and make any pre-start voltage level selections (e.g., 120 VAC phase to neutral)
3. Set ground rod, and run rod feeder tail back to genny safety ground Cam-Lok or buss bar connection (green).
4. Start engine and allow several minutes for warm-up prior to loading with the distro.
5. Adjust and meter hot phases to neutral for 120 VAC. Verify that neutral and ground have zero volts across them, with very low resistance across them and the genny chassis.
6. Connect your feeders to the genny output panel during warm-up in this sequence: ground, neutral and then hot feeders. Trip-off the genny breakers first if doing bare tails connections.
7. Make the final connections to your distro rack, and verify on the distro panel meters or by hand-metering that the correct voltages are present.
8. Place any mats, feeder anti-trip devices and yellow-jackets on feeder path to prevent accidental wire strain or damage.
9. Incrementally bring on your loading at the distro for the show.

After the show the process is somewhat reversed, with the priority that all the distro loads must be off before taking the generator off-line. Many thousands of dollars of equipment damage may result if a hasty genny operator keys off the engine with all the audio still live. Things like blown speakers, amplifiers and signal processors may become real if you hear the dreaded "bang" as the supply voltage crashes. Been there, witnessed that…

GENNY SELECTION AND OPERATION

When I speak of portable generators, I am not talking about those little hardware store, gas-powered units for camping, RVs or household power back-up systems. We are talking about trailered or truck-mounted diesel-powered generators — 40 kilowatts capacity or much larger. The rule of thumb is about a two-to-one ratio (or more) of genny capacity to distro maximum load capacity.

For example, a single-phase, 100 A-per-hot feeder leg could draw 24 kW (240 volts times 100 amperes). The reason for the genny derating is that music performances tend to be very dynamic in loading, and the generator must be able to handle low-load to full distro-load almost instantly. Generators use a slow reacting throttle mechanism to add more fuel to the engine in an attempt to keep the generator spinning at 1800 rpm for 60 Hz. Thankfully, the genny has a flywheel to store rotational energy for sudden demand changes, and a genny that's been properly lightly-loaded will handle most show dynamics.

Some constant loading will help in many generator usage scenarios. With stage lighting, the lighting console operator can be an ally by keeping the stage lighting more continuous, or creating a "load bank" of backstage always-on light fixtures to reduce the min/max loading range of the generator. In live sound, audio loading will be mostly at the mercy of performers.

When working with the show promoter, insist on helping with generator provider selection and generator type, or be specific on your contract riders. Strive to find a reputable genny provider with a good selection of "show-power" or "whisper-quiet" generators. Getting a low-bid, construction-grade generator should not be acceptable. Prevent these kinds of disasters before day-of-show.

We have all seen the Showtime listings in *FRONT of HOUSE* magazine listing "proprietary" when it comes portable power distribution description. Of the few who list non-proprietary distros, the list is typically limited to Motion Laboratories, J Custom Supply and a few others. While we're talking about power distribution we need to focus on portable power distribution Units (PPDUs), what constitutes the right and wrong way of building your own proprietary PPDU and how it should be operated.

IN A PICTURE

The picture shown in **Figure 13-1** is my 100 amp, single phase distro rack for live sound use. Brushing aside as much "proud papa" as I can, I want you to note a couple of non-obvious things. The first is the dual meters with green neon lamps above them that monitor the feeder hots before the main breaker. The next item is the obligatory "Electrical Hazard" sticker to warn people with half a brain or more to stay clear of this rack. These are not convenient "features" of my PPDU, but electrical code requirements for all PPDUs used for live performances.

Figure 13-1

While my panel meters may be slight overkill, at least the green tinted neon "pilot lamps," or some version of a pilot lamp, must be on the hot feeders, per section 520.53 of the 2003 National Electrical Code. I included the 250 VAC meters so that I can avoid hand metering after every hookup, and the 250 volt range is there in case of lost neutral that could push the meters into 240 VAC territory. Also, I chose neon indicators because they could handle the short-term over-voltage without burnout.

The Electrical Hazard sticker comes from several parts of NEC section 520 "Performance Areas," which states that only "qualified personnel" should operate PPDUs, and that these distros need to be "supervised" while energized and must be barred from access by the general public. I cannot speak for you, but I occasionally have nightmares of young kids playing "Tarzan" on my feeders dangling from a bull-switch. It only re-doubles my vigilance on keeping the distro rack in constant sight while mixing, making sure my feeders are strain relieved when making bull-switch or house panel tie-ins. My nightmare ends when junior Tarzan pulls out the feeder neutral and suddenly all my gear fries and select performers go into cardio-fibrillation when their mics and instruments become 120 volts above the stage floor. Let's hope the rest of you sleep more soundly.

BY THE NUMBERS

Building your own portable distro "pack" or "rack" does not have to be a nightmare, and can even be a source of pride when you do it right. Unfortunately, I see way too many questionable distros coming from people who should know better. So to help, I am going to skip through the distro specific sections of the NEC and hit the high points of the do's and don'ts of the code. Again, I cannot recommend getting your own copy, or at least temporary use, of the latest NEC book (from www.nfpacatalog.com) highly enough, and before you start you should review your grounding and wiring practices as well.

Section 518 of the code applies to Places of Assembly. It also describes a place of assembly as "drinking establishments" — you cannot get around the NEC just because you work in a nightclub or bar room, the code makes clear that these are included in its description of places of assembly. Just because you do not have a dedicated stage or "performance area" does not mean that your cabling practices and distro use can be done to loose residential standards. If you choose an electrician to assist you in PPDU construction, make sure they have plenty of commercial electrician experience, and don't just string romex in homes. This means they need to have experience with conduits, fittings and unsheathed wire — not wire staples and clamps for box entry. And dear God, leave the blue and grey plastic work boxes at the home improvement store. When in doubt, choose metal everything.

Section 520 applies to indoor performance areas. You may use indoor equipment outdoors as specified in section 520.10, but if rain occurs you must either immediately de-energize or apply a protective cover to comply with wet location requirements. I will say more on this in NEC section 525, regarding rules for carnivals and outdoor performances.

In section 520.44, distros are allowed a 50 % "diversity factor" on feeder wire ampacity. In other words, feeders rated at 100 amperes could actually feed distros of up to 200 amperes. This portion I am not comfortable with, but it presumes that usage is primarily stage lighting with partly dimmed fixtures, that fixtures are used for short times and that not all fixtures will be used at once. With the smaller 4 to 6 gauge feeder wires, actual performance loads should remain no greater than 80 % of ampacity. Most true "feeders" are considered 2 gauge and higher, and are not jacketed together.

In section 520.53 the NEC introduces the term "Road Show Connection Panel," which could mean a distro and portable stage light dimmer racks. The rules that apply are the usage of only metal enclosures, no exposed energized (live) parts, no concealed circuit breakers, stranded wiring only and pilot lights on feeder inlets that do not extinguish if the master breaker trips. This section also demands that there be no feeder allowed in trafficked areas, that there is twice neutral feeder ampacity in 3-phase systems and that the distros feature qualified personnel restrictions with markings on the panels (a.k.a hazard stickers). In this section there is also the stranded wiring clause which applies directly to feeder routing, but branch circuits still can use solid insulated wiring. You will find most 10 gauge and larger wires only available in stranded forms for commercial use.

Section 520.62 specifically addresses portable power distribution Units, but emphasizes mostly the types of cables to connect with PPDUs. All service cords (a.k.a. extension cords) must be "extra-hard service" of type S cords with heavy jacketing. All plugs, receptacles and receptacle breakouts must have appropriate strain reliefs (e.g., kellums grips). Consumer grade extension cords (type SJ or junior hard service) are allowed only if they are less than 20 feet in length, protected from traffic and less than or equal to 20 amp service.

Section 525 refers to carnivals, which means outdoor performances. Some of the obviousness that comes in this section is that no splices are permitted in feeders or other cabling, but what is not obvious is that all connectors that rest on the ground must be wet-listed — in other words, water-proof. That means all cables must be above ground at each end. No two or more extension cords are permitted to extend without elevated, wet-proof protection at the connectors is the rule. Above ground also applies to PPDUs. The section specifies this as at least 6 inches above earth surface contact. This mandate comes from the idea that sudden rainstorms could produce localized runoff of a few inches in torrential conditions.

The other major impact of the carnival section is that standard Edison receptacles for "general" use must be of a ground fault circuit interrupter (GFCI) type. What distro users must do is use the liberty portion of the "supervision" part of PPDUs and ensure the stage stringers that performers plug into have the GFCI protection. I keep GFCI breakouts just for this consideration. Other circuits can either be non-general twist-loc receptacles or standard Edison receptacles — but keep the performers from directly plugging into them. Experienced production crewmembers have all experienced GFCIs tripping from older guitar amps with oversized or leaky polarity capacitors, and sometimes giving a dedicated GFCI circuit to a problematic guitar amp is the solution.

THE BOTTOM LINE

Proprietary distros can be made to exceed the NEC provisions as well. For example, I used the more commonly available thermal breakers, but spaced them out to get less breaker-to-breaker contact in order to keep the trip thresholds reasonably high. Also, mounting in a standard shock-mount 10 U portable rack case fits right in with the amp racks in my system. And I use L14-30 receptacles for feeding amp racks, to make life easy using 4-wire 10-gauge sub-feeder patch cords. The L14-30 system of patching is also what Motion Laboratories sells in single-phase "Rack Pack" portable power distribution for amplifier and front of house racks. These are all benefits of Do-It-Yourself (DIY) distro construction for live sound applications.

I went through and gleaned the most applicable parts of the NEC that apply to use proprietary distro builders. What does not get emphasized enough is the "workman-like" distro construction practices. Neat wiring not only makes the system more inspectable, but also helps with the occasional fire marshal or safety official visits. City officials take visual cues, and feel assured that knowledgeable crew who meet the practice and the spirit of the regulations are on hand when the PPDUs are neat and professional looking. No flying wires, wire nuts or anything prone to vibrate apart from road use is more than a good idea. It also helps from a visual perspective to keep your distro and cabling clean and corrosion-free. Old distros can be useful, but keep them maintained and not looking like a public hazard.

When I wrote my first article on do-it-yourself (DIY) power distros, I received an overwhelming e-mail response. Based on that, *FOH* Editor Bill Evans and I decided that we needed to reprise the topic and answer readers' questions. Most of the questions were along the lines of: "Do you have plans for your distro?" or "Where do you get the meters, lamps and stickers" and "How did you connect your feeders to the panel, and what do you with the space behind the panel?" Well, lets answer those questions, introduce my club distro predecessor, and talk about some design issues.

THE VOLTS-WAGON

The picture in Chapter 13 **Figure 13-1** is my "Volts-Wagon" distro rack. I shamelessly pilfered the name from another distro on wheels, so that gives you lease to steal that name as well. I did not set out to make plans of the Volts-Wagon, so I do not have anything but pictures and my descriptions. If I did make plans and sold them, I could be in a heap of legal trouble should anyone construct a Volts-Wagon per plan and receive serious injury from a mis-wired version.

This first picture shown, **Figure 14-1**, is the back of the Volts-Wagon showing the four 2-gauge feeders exiting the rear bulkhead board and strain relieved by a conduit clamp with a red shop rag used as a clamp bushing. The feeders are 25 feet in length and terminate with 1016 Cam-Lok male connectors. This bundle of feeder resides within the rear of the Volts-Wagon when not in use. There is another 25 foot set of 2-gauge banded feeders (not shown) which also reside in the rear compartment. They feature 1016 female Cam-Lok connectors to bare tails for doing sub-panel or bull-switch tie-ins. Also not shown are the front and rear panels of the Volts-Wagon case. They have half pockets in them for stowing distro essentials such as ground and neutral (green and white) female-to-female Cam-Lok turn-arounds and a triple-tap grounding Cam-Lok "T" for genny connections. Also in the pockets are another 10 feet of 2-gauge feeder as a ground rod wire with a 1016 green male Cam-Lok for genny use, a digital multi-meter and various brands of 100 A, 240 VAC dual breakers for sub-panel tie-in (Square D, Siemens/ITE, GE and Cutler Hammer).

The Volts-Wagon is not a cheap DIY project. About $2000 went into it, and that is not including the two 50-foot feeder extensions that added another $900 to the bill. Most of the money went towards the road case and feeders. The circuit panel, breakers, work boxes, receptacles, panel meters and pilot lamps were the cheap part. The road case was done custom for me by David B. Little Acoustics and Woodworks (www.dblittlea-w.com), as he also made my 10-space shock-mount amp racks. The Volts-Wagon was meant to be identical in size to my amp racks so it would fit well in my trailer pack, and so that it would create a large even-height work surface if the amp racks and the Volts-Wagon were placed together.

The panel meters and pilot lamps were purchased through Mouser Electronics (www.mouser.com) as catalog parts. I bought the largest size 250 VAC meters available, but smaller meters can be also chosen. I purchased a few hazard stickers from Seton Industrial Supply (www.seton.com) for the breaker panels. These stickers also were placed at the rear of the amp racks, as both the power and speaker connections have enough voltage to hurt inquisitive fingers. The feeder and Cam-Loks were purchased from J Custom Supply of Baton Rouge Louisiana (www.jcustom.com) as built-to-order assemblies. It is best to let J Custom do the feeder banding and Cam-Lok attachment as they do it well, and connector attachment requires expensive special tools to do it correctly for reliable and water-proof connections.

Figure 14-1

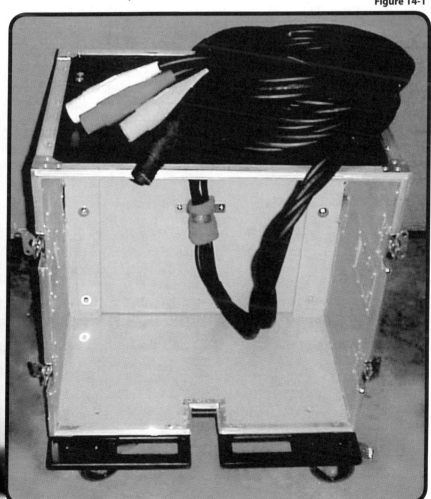

I have a couple tips to pass along for builders of DIY distros.

First, if you intend to have the feeders pass through the circuit breaker rear panel, choose a hole saw that is bigger for the panel and a slightly smaller diameter hole saw for the wood bulkhead. This ensures the feeders will not rub against the sharp metal panel box edge when exiting the breaker panel. The next tip is to borrow or invest in a set of Greenlee electrician's chassis punches, especially since the branch circuit holes may not be where you need them. Also consider a bottom rear slot on the road case to permit the feeders to exit the case with the rear panel cover closed. This helps keep wetness out and excess feeder inside. The rain use plan for the Volts-Wagon is to bag the case with a large clear poly bag for wet-proofing and to maintain visual status of service cords and panel lamps.

The Volts-Wagon design is based on the need to create a single-phase distro for my local soundco type gigs. Three NEMA L14-30 twist-loc receptacles were chosen for amp rack feeds for stage-right and stage-left mains amp racks behind the speaker stacks, plus another amp rack for monitor mixes. The 240 VAC, 30 ampere circuits permit the biggest sub-woofer amps to live on one 30 ampere, 120 VAC leg, with the mid- and high-amps sharing the other 30 ampere leg. Eight 20 ampere Edison duplex receptacle circuits are included to power all the other miscellaneous loads such as backline, front of house, stage lighting and fogger/hazer gear. A 50 A, 240 VAC range receptacle circuit was included just in case for extra lighting loads, or for being a kind soul to let another club distro feed off the Volts-Wagon's 100 A, 240 VAC capacity.

CLUB DISTRO

Based on the number of inquiries, many of my readers requested a good example for a smaller, club-capable distro system. **Figure 14-2** shows my well-worn club distro that was a predecessor to the Volts-Wagon. This club distro does not have a cute name attached to it, and I just refer to it as a distro-pack. The heart of the distro pack is a small Square D panel (Detroit Electric for us old farts) containing a 50 A, 240 VAC main breaker that is fed by a 15-foot feeder cable containing four 6-gauge wires, a kellums grip on the panel side and a NEMA 14-50P electric range plug on the other. The vast majority of clubs I work in have mating NEMA 14-50R range receptacles near the stage area. I also keep a 10-foot length of feeder cable terminated with a 14-50R to bare tails for the occasional sub-panel/bull-switch tap-in. Because not all performance areas have the juice close by, I also have 25- and 75-foot 6-4 feeder extensions equipped with 14-50 plugs and receptacles.

The club distro-pack feeds six 20 A circuits to Edison duplex receptacles. Originally, the distro-pack only had four 20 A circuits, but you may have noticed that the bottom two work-boxes (outlets) and middle breakers got modified to add more circuits. While I had to give up the idea of branch circuit panel lamps and full size breakers to make the mod, the extra two circuits added much more versatility. Again, as I noted in Chapter 13, the required hazard sticker and feeder panel lamps are present. Also, two panel meters are placed before the main breaker on the hot feeder wires to verify presence and correct voltages of the power.

You may have spotted the funky-shaped wood feet on the distro-pack. These flat-black painted legs of moisture-proof (green treated) plywood that are 12.75-inches tall to keep the bottom 6 inches off the ground, per NEC regulations for outdoor usage. Also, the legs perform other functions such as self-standing support when placed vertical to save space, or as spars when collecting the coiled feeder when transported. All six circuits on my distro-pack are Edison receptacles, so that means all my racks must be able to have power conditioning with Edison plugs for in-amp-rack distribution to L14-30 breakouts or direct to club distro-pack connects. Even with only six circuits this distro-pack can support backline, FOH, a couple modest amp racks and some stage lighting.

Figure 14-2

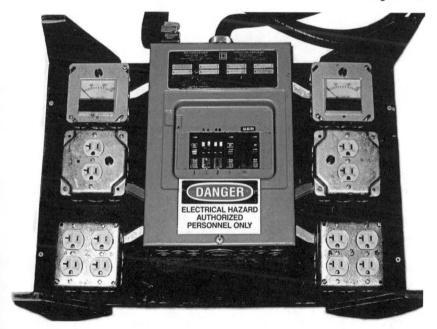

DESIGN THOUGHTS

When contemplating your next DIY power distro, brainstorm both the predicted loads (amp racks, lighting, etc.) and types of venues you will be tying into in the future. Regional and national soundco's will likely jump to three-phase 120 VAC distros with 200 ampere or more capability with 2/0 or better feeder sizes. Smaller regional and local soundco's will probably err on the side of single-phase distros so they can be flexible from nil power availability to using two phases of a three-phase power supply.

When choosing breaker panel and feeder ampacity types, do a bottoms-up calculation of the expected loads and how you want them to fit within the feeder busses. You will soon realize that a 100 ampere budget disappears like nothing when planning future racks and stacks systems. Be realistic in that each branch circuit will likely be used fully, but not all branch circuits will be used in each setup configuration. This is not like household power distribution, where you can de-rate feeder ampacity needs as a straight percentage of all the branch circuits added up.

A s you head into the outdoor gig season, a lot of concern revolves around scrounging for power for the show. Big shows can demand genny trailers and venue/city electricians for hookups, but many of the smaller performances are left to people who think any electrical access is enough to get the job done. You should be prepared to know what to do in a situation where there's not enough power.

POWER HOGS

When thinking about live sound and power consumption, audio power amplifiers are the first thing that pop into most people's mind. In reality, however, power amps reproducing live music do not consume much average power — the peak power demands are how most electrical power is allocated. This is easy to see with today's power amplifiers that are capable of 2000 watts per channel of audio power output, yet stay well within the 20 ampere, 120 VAC circuit capacity of 2400 watts. When you have tough power gigs ahead of you, how do you choose power amplifiers that can be efficient?

Power amplifiers typically break down into two sections, with a couple of variants per section. In modern amps there is the amplifier section and the power supply unit (PSU) section. Within the amplifier section the two major design categories are linear and switcher. In the linear category the popular high power amplifiers break into classes of AB, G and H. In the switcher category the classes are D, I and TD. In class AB, the most efficiency you can get is about 66% at full output, with the remaining 33% or more lost as heat. Classes G and H extend the AB circuit design with two or more steps of increasing power supply voltages that engage as the input signal increases. As a result, amplifier efficiency can increase to about 80%, not bad when you consider heat management to circuit complexity.

Digital power amps in the D, I and TD classes typically can render efficiencies up to 90 to 95%, permitting huge output powers in very small rackspace packaging. So if you have efficiency on the priority list class G and H linear amplifiers plus all the switcher amplifiers are candidates for sipping power.

On the PSU section front, conventional power supply units use traditional 60 Hertz transformers to match the inputted 120 VAC to the required DC voltage rails of the amplifiers. Of course, high power capable transformers have a lot in common with bowling balls, especially weight and size. Switcher PSUs chop the incoming 60 Hertz 120 VAC at high frequencies. And at these 80 kHz to 500 kHz switching frequencies, the transformers shrink in size and weight to small palm size devices while moving the same amount of power as conventional 60 Hz transformers. Think of your amplifier circuits being re-supplied with energy at 250,000 times a second, instead of 60 or 120 times a second. The re-supply amounts can be smaller and more responsive to varying energy demands.

But switcher PSUs have more complexity compared to the brute force of conventional transformers. So issues of size, reliability, weight and cost have to be compared when selecting a power amplifier. Both conventional and switcher PSUs have about the same efficiency, so that does not help or hurt the inadequate power problem. But for us humping amp racks, the light weight and smaller size of switcher PSUs win out over the cheaper, heavier, bigger conventional power supplies.

THINK STACKS

So now that you are biased towards efficient amplifiers, don't forget that less speakers require less amplifiers and less power to drive them. Most low frequency drivers convert nearly 80% of the power at the voice coils to heat. So in order to maximize your gains out of the remaining 20% power that can be used for acoustic output, make sure you have more efficient speakers, to use less of them. Things like horn-loaded mids and lows are not just remnants of the Woodstock generation, a lot of modern touring speaker stacks still retain the horn-load design for efficiency. You can refresh your brain on these thoughts in the section on speakers and amps.

OTHER LOADS

Backline power for musos (musicians) is something that cannot be overlooked when conserving electrical power draw. While many groups can get by with a single circuit, you may be fighting a lost cause when the Marshall stacks and Ampeg SVTs invade the stage. If possible, communicate with the artists ahead of time about the power shortage, and persuade them that smaller backline gear will make the show a go instead of a circuit breaker popping event. Also, let them know that the 700 to 1200 watt fogger/hazer machines will not be permitted for the gig.

Thankfully, the consoles and processing at front of house and monitor beach tend to sip power, and in some cases these loads can survive being paired up with other audio loads. Also ripe for pruning are the number of monitor mixes and the redundancy of wedges per performer. Each wedge sucks juice, and that can add up fast. Also, note that larger digital consoles are tending to be power hogs as well. That's something to consider when thinking about a 300-watt analog console with 200 watts of outboard signal processing versus a 1000-watt digital console and zero outboard processing.

One tip I wish to share is to begin placing power conditioner/monitor devices in your racks to keep a visual indication of current draw. I prefer the older Furman PM-PRO and PL-PRO power conditioners on racks, with the AC draw represented in a LED bargraph, so that I can monitor from a distance the average and dynamic power consumption of each circuit. I can't do that with subwoofer amplifiers, but it helps to know your limits on other circuits monitored.

SNIFFING OUT POWER

When you are giving up on using your power distro, and are down to the detective work of finding separate 20 ampere venue circuits, you have a few tools to use. The classic tool is a neon bulb circuit tester or a plain old night-light. You snap off selected breakers and track down dead receptacles, and in this way you can map out a branch circuit. Of course, this takes precious time.

Another method involves an electrician's tool called a circuit sniffer or breaker finder. Knowing where the venue's breaker panel is, and with a little electrician intuition, you can make pretty good guesses on the circuit paths on strings of receptacles using the sniffer. Traditionally, a circuit sniffer works by sticking a transmitter device in a energized receptacle, and placing the sniffer device on the breaker that receives the transmitter tones the loudest. But sniffers can also be used to sense the receptacles between the transmit receptacle and the breaker. So the clever choice of the transmitter location can buzz out common circuit receptacles without having to make treks back and forth to the breaker panel.

You really want to ensure that all your circuits run to the same panel, at least on the audio loads. I do not recommend using a different panel feed for stage lighting loads, but you may be okay if both panels join to a master panel. When in doubt, meter your grounds and neutrals between differing panel circuits. In most newer commercial buildings, you may have success with this, but watch out for old venues that could have multiple master power feeds due to decades of remodeling. Also consider having the new non-contact 120 VAC finders that use your hand as the safety ground reference. These finders are typically the Fluke Volt-Alert or Triplett Cricket that light up or chirp when they are in proximity of energized wires. Ideally, all grounds and neutrals will not indicate, only hot wires will indicate. Volt-Alerts and Crickets are also great for finding disconnected cords when you expect them to be energized.

LOW POWER OPERATION

Firing up a live sound system under constrained power conditions requires an easy-does-it approach. First on the checklist should be turning gear on separately instead of a hasty run to the amp rack and toggling power switches. During the actual gig, look for ways to get the loudness good enough, but quieter than usual operation. The same goes with stage light operation, as it can also topple audio circuits by heating nearby circuit breakers. Keep the energized fixture count on scenes low, and attempt to switch scenes incrementally through fades instead of hard ons and offs. All those sudden stage light power-ons have nasty peak current draws that can push a breaker into the trip threshold.

The science of household and commercial circuit breakers bears repeating. These breakers trip through a thermal means where a small (shunt) resistance is monitored, and if hot enough, will release the branch circuit from the main buss of the panel. Minor overloads will take minutes to trip, but major overloads will trip in seconds. But the important fact is that a warm-to-hot breaker trips quicker if the load is not lightened. And a hot breaker also warms adjacent breakers, further lowering their trip thresholds.

"**O**f all the mail that I receive, the subject of power distribution seems to bring the most pleas for more words on the subject. I have plenty of stories of arcs and sparks, but thankfully few have resulted from my handiwork. I will tell you a secret, I have been deathly afraid of medium voltage (120, 240 volts) ever since I was a kid, and I think that my healthy respect has helped me from being over-confident. Every now and then my parents remind me that as a toddler I would stick my hands into uncovered outlet boxes, receive a shock, run crying back to my parents — and then repeat my curiosity a few more times that day. And you wonder why I am an electrical engineer today?

But this chapter is about power distro connections, and the dos and don'ts of making them. I am going to let the pictures do the talking, but I will sprinkle in a few anecdotes to explain the thinking.

CAM-LOKS

Throughout my 30-plus years of being a musician, it has only been in the last 10 years that I found out what Cam-Lok connectors are, and how they are used. Most of you have seen them at carnivals and fairs but never thought they would become part of live sound production. Most high power feeder wires use these single conductor, insert and twist connectors for 8-gauge wire and larger feeders. **Figure 16-1** shows a banded and coiled group of 2-gauge feeder wire with 1016 Cam-Lok male and female ends for 100 amp, single-phase service.

Cam-Lok connectors come in three sizes (1015, 1016 and 1017), with size 1016, with a maximum rating of 400 amperes, the most common. The size 1015 connectors are typically dubbed mini-cams for 8-to 4-gauge wire and a maximum rating of 150 amperes. I rarely see mini-cams in usage, but when I do they are usually with stage lighting dimmer racks and some touring power amp racks. Even rarer are the sighting of 1017 maxi-cams with 350

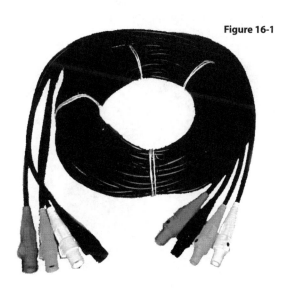

Figure 16-1

to 750 mcm feeders and a maximum rating of 690 amperes. The usual 1016 cam-loks can handle 2 to 4-0 gauge feeder wire.

Most Cam-Loks are weather-proof, with their colored rubber boots protecting from innocent fingers when open, and forming a water-tight connection when connected. Cam-Loks also come in panel-mount as well as cable connectors. When dealing with Cam-Lok feeders (carney cable) and Cam-Loks, it is best to let a trained person with the right fixtures and jigs do the connector to feeder attachment to ensure a tight seal on the boot to the cable jacket.

STRAIGHT-BLADE PLUGS AND RECEPTACLES

Of the lower amperage connectors the straight-blade dryer and electric range plugs win the popularity contest due to lowest cost and availability. Yeah, many touring racks have 30 and 50 ampere twist-lock power interconnects, but there is almost a hundred dollars of brass and nylon involved in each pair of plug and receptacle. For us weekend warriors in the clubs, the presence of a NEMA 14-50R range receptacle at the back of the stage means that we are plug and play, instead of plug and pray for the illegal receptacles that force the neutral and safety ground onto a single contact.

Of course, not all receptacles are wired 100 % correctly. That is why you should have a working knowledge of the commonly encountered receptacles, and know which slots are supposed to be hot, which are supposed to be neutrals and which are grounds. And do not leave it at metering the 120 volts to neutral, but check for no voltage between the safety ground and neutral contacts — then check the resistance between safety ground and neutral for decent continuity. To get to know what NEMA 14-50 range plugs and receptacles look like, **Figure 16-2** shows the pair.

Everyone should be familiar with the NEMA 5-15 household "Edison" plugs and receptacles, but the most common good distro plugs and receptacles are the four-contact NEMA 14-30 dryer and NEMA 14-50 range devices. The older

Figure 16-2

NEMA 10-30 dryer and 10-50 range connector with the 3-prongs (hot-hot-ground), were intended for the older 240 VAC-only appliances, and were grandfathered out of usage back in the 1980s. But I still see old local sound-co's with 10-50 plug adapters on their club distros cheating death and lawsuits by using these older club receptacles. One soundco — which shall remain nameless — even added a green-wire spring clamp to a 10-50 plug under the deluded idea that the club's feeder conduit might actually be a decent safety ground path. **Figure 16-3** shows three NEMA plugs, the 5-50P (illegal), the 14-50P (legal) and the 10-50P (illegal).

Figure 16-3

TAILS

Distro tails are typical moderately short pieces of feeder cabling that have a receptacle (female connector) at one end, and 6 to 12 inches of loose conductors at the other end. I have 10- to 15-foot tails for both my Cam-Lok feeders (2-gauge) and my 14-50 feeders (6-gauge). Most tails shrink with usage as the loose ends are bared a half-inch for screw termination into bull switches, circuit breakers or ground or neutral buss-bars. Be wise and always de-energize your hots before connecting the bare tail ends into the screw orifices. As your bare stranded copper wire ends fray with usage, trim back the frays and strip new ends. And, as I have to state: If you are not a licensed electrician, do not play with tails, or the legal professionals may lay waste to your financial health for the foreseeable future. While most moderate to bigger clubs do provide distro receptacles, some may only have bull switch/sub-panel breakouts available. **Figure 16-4** shows my 6-gauge set of tail loose ends with freshly stripped bare copper wires. Note that I did not pull all the feeder insulation off the ends, but left it in place to prevent fraying before connecting.

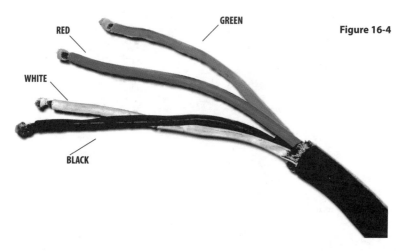

RED

GREEN

Figure 16-4

WHITE

BLACK

SIGNAL PROCESSING

This signal
processing section is a
collection of all little details
of getting electrical audio signals
from the source to end of the signal
chain. As an electrical engineer, I may
overemphasize the details of signal
processing, but the nuances of signal
processing mean the difference
between adequate sound
reinforcement and a show
you really enjoy hearing.

"Out of Phase" sometimes really gets on my nerves as a catch phrase. To me it is like a stopped clock being right twice a day, or once a day on military clocks. Theoretically, a signal is out of phase if it absolutely does not match up to its reference signal. And we are talking about 0.000001 degrees here.

Then there is polarity. Polarity is as binary as it gets. Either the signal is an absolute copy of the reference signal, or it is an inverted copy of the reference. So what is the confusion? The devil is in the details.

BALANCING SIGNALS

Unbalanced signals are thought of as just the signal "hot," containing the normal or "reference" polarity, and a ground return that provides a Direct Current (DC) voltage reference. Balanced signals contain hot and "cold" signal lines, plus the ground return. In most professional systems, the cold signal is a mirror image or an opposite polarity signal of the hot signal. These are NOT called in-phase and out-of-phase signals, as both signals are in-phase but opposite polarity.

Balanced signals have at least two advantages. The first is that most receiving circuitry gets twice the signal voltage amplitude (+6 dB) because the circuits differentially measure the signals instead of taking from a ground reference level. The second is that any noise added from circuit transmission to reception tends to effect both the hot and cold signal lines equally, and is therefore cancelled out. The ability for balanced signal receive circuits to reject in-phase noise signals is called "common-mode rejection."

In my experience, one of the biggest mistakes committed by new persons assembling sound systems is using unbalanced over balanced signal connections in critical signal paths. I personally have seen systems that had the power distro ground wire left open ("flying") in the breaker panel to resolve a buzzing noise problem. Never do this when an XLR patch cable is such a simple fix.

DEPICT THIS

Figure 17-1 shows three signal plots showing signals of differing polarity or phase shift. I deliberately choose a more "real world" signal type than showing nice clean sinusoidal squiggles. These unclean signals represent signal sources with a fundamental wave and other summed waves that may be harmonics or other signals that got mixed together. The point being, these signals may have non-symmetrical amplitudes, and positive and negative peaks.

Trace A is used as the example of normal or reference polarity, and two cycles are shown for brevity. Trace B is the exact same signal, but in the opposite polarity. From this you can visually see why a voltmeter placed across the traces

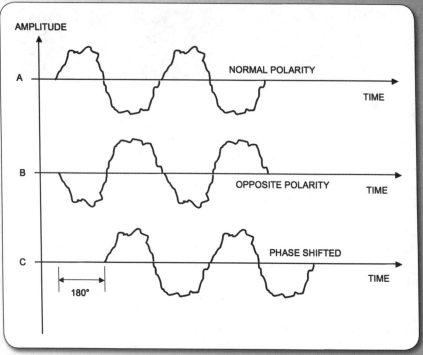

AMPLITUDE

A — NORMAL POLARITY
TIME

B — OPPOSITE POLARITY
TIME

C — PHASE SHIFTED
TIME

180°

Figure 17-1

(trace A = hot, trace B = cold) would provide double the signal amplitude. Trace C represents a time-shifted version of Trace A with about 180 degrees of phase shift with respect to the traces. Note that Traces B and C do not resemble each other because of the non-symmetry of the signal.

EXAMPLES

To help reinforce the point, common polarity examples would be similar to drum head miking, where a mic placed above the head would yield an opposite polarity signal than one placed below the head. Polarity switches on older guitar amplifiers are also found to select which AC wire best represents a ground reference to the amp's chassis. Hopefully, this is an unbalanced power distro supply with the wire selected being the neutral and a near ground reference.

Polarity switches also show up in channel inputs of most mixing consoles to handle the drum-head example of miking. If two mics (above and below) are used in this example, one mixer mic channel should be polarity flipped to provide a summation of the drum's signal.

Phase-shifted signal examples are easy to find in live sound environments. Anything miked in air with different source distances enjoys the singular fact that acoustic waves will travel at the speed of sound (nominally 366 meters per second), while electrons through wires travel closer to 30,000,000 meters per second. Thus, electro-acoustic transducers (speakers, mics) send or receive phase-shifted signals, once your ears become the reference point. This is why dual-miking, speaker arrays and bi-amping become the focus of phase correction methods, not just polarity.

Room acoustics and reverberation are the classic example of the listener hearing their "reference" signal from the closest speaker cabinet near them. All other speaker cabinets and room reflections are delay versions of the reference. At best, they can be called room ambience — at worst, they decrease signal intelligibility to garble.

Any signal that is "filtered" will have a phase shift relative to its unfiltered reference. EQ strips, crossovers, DSPs and even DC-blocking circuits will induce modest phase shifts. These are all time delays that convey degrees of phase shift, depending on the fundamental frequency. The real issue is how much phase shift and if there are other un-filtered versions of the reference that may be combined down the signal path.

TALKING POINTS

The key points to bring home are that signal polarity connotes no change in phase, but only normal or opposite (inverted) signal forms, and phase shifted signals have electronic components or acoustics employed to slow them from the reference signal. Finally, while a pure sinusoidic signal shifted 180 degrees may be truly looking like an opposite polarity signal, such examples are fairly rare in practice.

PHASERS, FLANGERS AND CHORUSES

I want to comment on these phase shifting devices because while they all use similar principles, the sounds they produce can be radically different. Phase shifters employ cascaded resistor-capacitor networks to vary the resulting time-delay, creating comb filter effects — much like having a several synchronized wah-wah pedals operated at once. These comb-filter effects are the result of using the reference signal against a varying phase and amplitude effected signal. Hopefully, you choose to have a phase effect, and it's not just a result of having have your sound system being a permanent version.

Flangers borrow from the recording studio trick of two near identical sources tracked on separate reel-to-reel decks with the operator placing his or her thumb on one of the reel flanges to delay the playback of the second source. Today, dedicated circuits perform this trick without the mechanical intervention. Choruses are sister elements of flangers, in that delays are created, but the sampling method of choruses permits controlled pitch-shifting effects when the delays are being modulated in time.

Well, I couldn't put it off any longer — this analog curmudgeon has to acknowledge that the invasion of digitized live audio has begun, and it will benefit us all. I spent many hours diving through the jargon and FAQs to bring you all the dirt on the Ethernet and the popular live sound network schemes. I tell you, my head hurts — and now I understand why IT networking professionals command six-figure salaries.

No, I am not a newbie at digital and the Ethernet. I have done my penance designing communications protocols, and seen enough ack/nack packet ladder charts for a couple of lifetimes. My challenge to you is to kick back and digest the fire hose of information I am about to lay on you. I will attempt some layman descriptions along the way.

THE ETHERNET

I am going to forego the Ethernet history lesson and mention that the Ethernet is a seven-layer system of protocols that define the method of network communication. These layers are briefly described in **Table 18-1**. A protocol is simply a set of rules that narrow the freedom of how data/signals flow. It you relate the Ethernet to using a highway system, the physical protocol would be constraints on the road and lane sizes, the min and max speed limits, the amount of vehicles that could pass per time and the sizes of the vehicles. The analogy applies to the datalink layer as the vehicles have to have defined license plates, bumpers, tires, weight restrictions and a definite start location and destination addresses. Extending this to the network layer, you would describe the payload (human or cargo), how many items and the order/method of unloading them.

Two dominant Ethernet schemes are used today in live sound: CobraNet and EtherSound. While I plan to hit the high points of each, the common physical layer aspects each need to be explained first. Both use a cable system with RJ-45 8-pin connections with at least four wires (two twisted pairs) defined as 100 Base TX, or capable of 100 Megabits per second transmission rates up to 100 meters. As far as the manufacturers are concerned, they would be thrilled if this were the end of the technology discussion, and we just bought what we needed from here on out.

Table 18-1 ISO 7-Layer Protocol Description for the Ethernet (IEEE 802.3)

	Layer Title	Description
Layer 1	Physical	Hardware (RJ-45, 100Base TX, etc.)
Layer 2	Data Link	Data Frames and their Transmission (collision detection, half-duplex/full duplex)(primary firewall)
Layer 3	Network	Addressing, Data Packet Formatting and Routing (internet protocol (IP), packet routing)
Layer 4	Transport	Reliable Data Transfer (TCP or UDP, client/server handshake, ack/nack)
Layer 5	Session	Communication With Remote System (starting/stopping communications)
Layer 6	Presentation	Data Representation (data presentation to user)
Layer 7	Application	Applications Using the Network (smtp, ftp, http, dns...)(secondary firewall)

MA BELL TO THE RESCUE

I want to spend some quality time on the cable and connector description, as it will pay dividends in related ways. The whole "RJ" system of connectors was developed for low cost deployment, but needs special attention as it is the weakest link in a network. When dealing with RJ connector cable wiring — described as category 5 unshielded twisted pair, or Cat5 UTP — the color code is borrowed from the telephone industry. Telephone cable wire basic colors in order are (1)Blue, (2)Orange, (3)Green, (4)Brown and (5)Slate(gray). Additionally, each wire is striped and has another color called a binder color that indicates a multiple of the five wire color groups, which all adds up to 25 circuit bundles. The binder colors are (1-5)White, (6-10)Red, (11-15)Black, (16-20)Yellow and (21-25)Violet. For most RJ-45 wiring the binder color will always be white, and blue, orange, green and brown as basic colors for the eight pins.

To complicate things a touch, all wires come in pairs with tip(+) conductor using the binder color on the insulation, striped with the basic color. The ring(-) conductor is twisted with tip conductor and is reverse colored, meaning the basic color is the insulation striped with the binder color. So for example, pair (1) is white/blue (tip) and blue/white(ring). Note that there is no sleeve or ground reference in this cable system. Each of the conductors is usually a flimsy 24-gauge solid wire in the Cat5 cable, and prone to break if flexed improperly. Note that Ma Bell wires to most homes with 19 or 22-gauge pairs, with two pairs provided in case of breakage or if a second phone service is later required. Also note that the blue/white and orange/white pairs are used up to your Network Interface Box on the outside of your residence. With Ethernet wiring, you may find yourself cross-patching to good pairs of wires on RJ-45 connections should you not have spare Cat5 cabling.

A quick note should be made that in the RJ-45 jack pair(1) is usually left spare and sometimes wired. Pair(2) (orange/white) is the DTE transmit data pair and pair(3) (green/white) is the DTE receive data pair. Pair(4) (brown/white) is usually not wired. All end-point gear is defined as DTE (data terminating equipment) and can be considered the senders and receivers of the digital audio traffic. DCE is data communications equipment and can be considered gear that combines, splits or filters (i.e., a firewall) data between DTE elements. DCE are hubs, routers, servers, gateways, bridges and switches to pass data to the audio handling processors (patch panels, consoles, computers, etc.). Imagine a data highway with two lanes with traffic going in opposite directions per lane (full-duplex, transmit and receive). Even though some digital audio DTE are dedicated senders or receivers, each has a return path to retrieve commands or status information from "master" DTE.

SPOKES AND CHAINS

Getting back to CobraNet and EtherSound networks, each has its own recommended wiring methods. Cobranet's preferred method is a hub and spoke interconnection system where all DTEs connect to a single piece of gear, called an Ethernet hub, no matter if you are primarily a sender (master) or receiver (slave) of digital audio. One DTE element is defined as the primary master, or "conductor," and sends about 750 "beat" frames per second as means of sequencing up to 64 channels digital audio down the transmit wires to the hub. The hub takes all the DTE transmits and sums them back as receive data to all the DTE receive wires. Unlike normal computer Ethernet protocols, Cobranet sets up a synchronus (isosynchronus) form of frame traffic where the conductor DTE orchestrates the sequence of other DTEs to transmit their digital audio data. Typical Ethernet non-audio purposes just send frames of data from the DTEs (PCs) whenever they feel like it (asynchronus), and any accidental collosions are retransmitted after a random delay.

EtherSound has a preferred method of chaining DTEs along a one-way path, much like you would do with a MIDI system. EtherSound DTEs usually have two to more RJ-45 connections with receive (in) and transmit (out) jacks denoted on each DTE element. DCE elements like hubs can still be used, but only if all the DTE transmitters are chained upstream on one spoke, and the hub multicasts to multiple receive DTE elements (like main and monitor digital consoles). The advantage of EtherSound is that chaining removes the hub, and the 64-channels of 24-bit, 48k-sample digital audio flow in only one direction from senders to receivers. Digigram is the parent of the EtherSound network protocols, and also manufactures single rackspace DTE patch interfaces such as the ES8in for eight XLR jacked line level signal inputs, ES8out for eight XLR analog audio line outputs and an ES8mic for eight XLR mic-level inputs. Thus, one could design sub-snake rack cases with multiple ES8mic and ES8in boxes chained back to monitor beach's console, then run over Cat5 up to the FOH console. Then each console could send Cat5 cables back to side-stage ES8out boxes for main and monitor mixes or directly to EtherSound capable speaker processors or power amplifiers.

LESS OF THE CABLE MESS

While both CobraNet and EtherSound will have at least two networks (digital audio sources to consoles, digital audio mixes from consoles), both share the isosynchronus traffic requirement to minimize audio latency, and both have slightly different control methods at the datalink and network layers. While CobraNet may look more difficult with spoked hub setup, if the sender and receiver DTE are just two boxes, no hub is necessary, and all 64-channels are available between the two DTEs. An example of this would be a single snakebox serving as the sender DTE and a single digital console serving as the receiver DTE and conductor element, plus doing both main and monitor mixes to a return CobraNet. Do not be afraid of acquiring computer-store off-the-shelf Ethernet hubs for CobraNet, as their reliability is proven every day by cubicle PC rats like us. **Figures 18-2** and **18-3** show typical CobraNet and EtherSound networks.

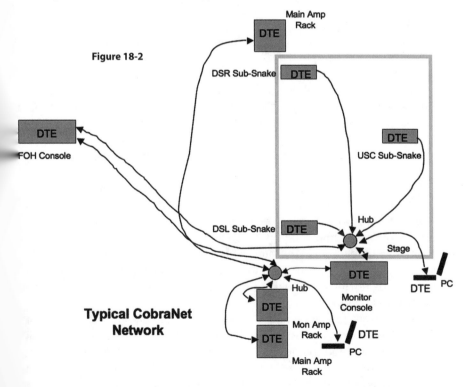

Figure 18-2

Typical CobraNet Network

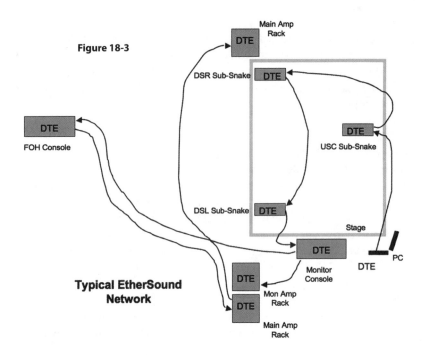

Figure 18-3

Typical EtherSound Network

CONSTRAST AND CONCLUSIONS

EtherSound uses the above-mentioned chain method of moving 64-channels of 24-bit, 48k-sample digital audio around to places of need. Companies like Innovason, Fostex, Digigram, and others have adopted Ethersound as their system of choice. CobraNet uses a spoked method for 64-channels of 48k-sample digital audio with selectable resolutions of 16, 20 or 24-bit digitization. Companies like the Harman Group (JBL, Crown, BSS, Soundcraft, etc.), QSC, Peavey and others have adopted CobraNet as their system of choice. Ether-Sound has a fixed latency of 6 samples, and CobraNet has a variable latency of up to 256 samples.

The main thing to remember about live audio using the Ethernet is that each bit is transmitted at 10 nanoseconds, while each audio sample (usually 3 bytes) is moving in 0.24 micro-seconds, excluding the frame and packet overheads. Moving 64 samples brings the math up to about 15 micro-seconds, leaving about 90 % or more of the latency budget for overhead and DTE asynchronus messaging. With latencies in the fraction of a millisecond to a couple of milliseconds, no human should be able to detect any delay in the performance due to the networks.

Lastly, I want to extend kudos to Neutrik for coming out with a series of rugged RJ-45 connectors using XLR-like features for Ethernet cabling. Now all we need is some tour-grade Cat5 UTP for 100 Base TX transmissions. Ethernet connections are now possible in fiber-optic (100 Base FX), and will extend transmissions from 100 meters to 2 kilometers using media converters.

One of the tricks a live sound person has to make the mix sound better than what it is coming off the stage is the selective use of compression. However, let me jump up on my soap box to preach that compressors are not a cure-all for other production and performance deficiencies. The thing that irks me the most about newbies in this business is that they feel the need to have at least one dual-channel compressor, just because they see the pros using them in their FOH racks. Of course, not knowing where a compressor is inserted, they throw it across the main mix and "tweak it in."

What entry-level sound persons need to remember is that compressors come last on the list of procurement priorities. If you wasted $200 on some compressors, that money could have bought you the next size larger mains power amp, or better mains speakers that could really make your production stand out. Living with inadequate racks and stacks and squashing your mix with a compressor to make up for that inadequacy is something that infuriates me. I am now stepping off my soapbox.

COMPRESSOR GUTS

Before we jump into where and how to use compressors, let's review the internal functions and how a compressor's controls work. Compressors have two paths for the signal once it's buffered at the input. The main signal path is pushed into a Voltage Controlled Amplifier/Attenuator (VCA) stage. The controlling voltage for the VCA comes from second path, or "sidechain," that takes the input signal, converts it to a logarithmic representation (dBs), threshold detects the signal, and gains the signal to the VCA to amplify or attenuate control inputs. Audio signals exiting the VCA are fed to a last stage for "make-up gain" if the user desires. Main compressor controls such as the threshold and compression ratio work upon the sidechain signal to set the points of compression onset and the rate of compression.

Additional controls like attack and release work on the sidechain signals between the threshold and ratio controls to place smoothing on the signal as it passes the threshold and returns from the signal peak amplitudes. Many compressor manufacturers offer an auto-attack/release feature that continually adjusts the attack and release time constants based on the rate of change of the incoming sidechain signal. If you are not looking for a specific attack or release time constant, or just do not know the best settings, consider switching in the auto-attack/release function.

Another optional control that sometimes appears on compressors is the gate/expander control. Mostly used for recording, the gate expander is just another dynamics alteration in which low-level buzzes, hums or non-musical noises can be un-amplified until gating signal amplitude (threshold) has been reached. I personally leave the gates off for live sound use, due to my own uncertainty that I will remove the set gating at the end of the gig, and next gig I am in the midst of level-checks and can not find the channel's signal until I have dialed in way too much gain. These sudden blasts of channel levels are unwelcome to my PFL headphones and my ears, and I can live with a little more noise floor to prevent the occurrence.

SETTINGS

The majority of compressor channels are used as channel inserts for key signal types with sometimes nasty dynamic signal levels. Mostly inputs like vocals, bass guitar, keyboards, brass instruments and some percussion. If you use insert compression, you may want to bypass the compressor or raise the compression threshold way up when doing line checks. Also, you want to set the channels a bit lower than your normal levels, such as 0 dB for average levels. The rationale for this is the nominal +20 dBu clip level at the channel preamp that gives you the 20 dB gain reduction headroom before clipping. For example, a 4-to-1 compressor ratio at a 0 dBu threshold brings the maximum compressor output level to +5 dBu before clipping.

Compressor make-up gain settings on insert channel compression should be used sparingly, if at all. Gain settings from 0 dB to +3 dB are typical and prudent. The rationale for this is that occasionally you may have to un-insert or bypass the insert jack signal processing, and the sudden drop in level may not be something you can tolerate in mid-performance. Generally, your channel fader is your best tool for channel level adjustments. If you find yourself often tweaking preamp gains or compressor makeup gains, you probably have a gain structure problem best addressed with channel, subgroup and master faders.

VOCALS

The biggest payoff in compressor usage is with insert compression on lead vocals. In general, the compression ratio for live performance lead vocals varies between 2-to-1 and 4-to-1, with the threshold set for about 6 to 10 dB gain reduction on the peak vocal signals. Most kindler, gentler music genres err towards the 2-to-1 side, with rock vocals towards 3-to-1 or 4-to-1. If your compressor has the hard-knee/soft-knee threshold switch, choose the soft-knee for vocals. The effect of compression on vocals is best described as a "more professional sound," due in part that most of us hear plenty of compressed vocals on television and radio broadcasts and in recorded music.

Backing vocal compression is nice touch, especially if you have the luxury of plenty of compressor channels. You can use the same kinds of settings as lead vocalists, and the compression helps keep the backing vocals from crowding the lead vocal. In a pinch, some compressor-poor sound persons will combine all the backing vocals to a subgroup, and use a subgroup insert to compress the vocals together. This can work well if the backing vocalists live on the same monitor mix or are bunched close enough together to hear each other's vocal levels and self-adjust their voices. Things can get vicious when sub-group compressed vocals get dominated by one singer, who then causes the compressor to focus on that voice, and thus squashes the other singers further down into the mix.

BASS GUITAR

Insert compression on bass guitar direct-inject (DI) signals is the next natural choice on compressor channel allocation. Much depends on the style of the bassists. For example Jazz bassists may not need any squashing at all — and on other side, a thumb slappin' and string pullin' bassist may never get enough dynamics control. Bass guitar signals typically get a 5-to-1 to 10-to-1 ratio and gain reduction in the 10 to 15 dB range. This is especially handy to smooth up less talented bassists, or bassists who alternate between a plucking and slapping technique. A hard-knee threshold setting is appropriate here.

KEYBOARDS

Some keyboards, especially synthesizer keyboards, are ripe for compression. Not so much for the performing style of the instrumentalist, but because from synth-patch to another patch the levels may vary widely. Here the compressor is set up more like a limiter, with a higher ratio such as 6-to-1 to 10-to-1. The softer patches may not see any gain reduction and the louder patches should be adjusted for no more than 15 dB of gain reduction. Also, using newer and nicer multi-band compressors (e.g., TC Electronic Triple-C) for low/mid/high frequency bands may help, especially in sequenced multi-part synth playback. With keyboards, a soft-knee threshold helps soften the gain reduction onset.

KICK DRUM

Optionally, the kick drum may get a taste of compression if a "punchy" kick drum sound is desired. Most percussion is generally not compressed, but some may have compressors set up as limiters with thresholds at +10 dBu and ratios at 10-to-1 or higher. Kick drum punch is usually dialed in with 4-to-1 to 10-to-1 ratios and thresholds set to taste, typically in the 6 to 15 dB of gain reduction. The hard/soft-knee switch can be set arbitrarily to whatever appeals to the user.

BRASS

Lastly, brass instruments are another compressor application use — like keyboards — to strangle direct hits on the brass microphones, especially when the instrumentalists tend to be dancing to the groove. Setting a compressor channel with 8-to-1 or higher ratios, and 10 dB or more of gain reduction on the peaks helps keep distortion on the brass channels from getting to be a problem.

SQUASHING THOUGHTS

While I do not expect everyone to have a dozen compressor channels, it is nice to have at least a few to handle the critical offenders of excessive dynamic range. A typical soundco will usually have one or two premium ("diva") compressors for the star vocals and a bunch of general purpose dual or quad channel compressors. I find that having a couple quality name compressors for vocals plus a couple adequate compressors for instruments works well in most gigs. One thing to avoid are compressors that use external power supplies such as wall-warts and line-lumps, as they generally do not provide the needed +18 to +22 dBu headroom for live sound mixing.

Good uses for compressors placed on console outputs are generally few, but sometimes warranted. In the chapter on ringing out a room I mention comps for catch and hold ring-out tactics on aux send masters. You can also use compressors as limiters when the downstream amplifiers are without clip limiters or too large for protecting the mains speaker drivers. Thankfully, the emergence of DSP speaker processors with onboard limiters is beginning change things.

Now that Digital Signal Processors (DSP) have made their way into stand alone equalizers and speaker processors, a sound engineer needs to have an intuitive sense of parametric filtering when smoothing out the bumps of system responses. While things like shelving filters and crossover points are easy to visualize on a scale of audio frequencies, the classic peaks and dips of parametric bell filters are more difficult. The following will provide a refresher of the basics and a handy chart of bandpass filter widths.

PARAMETERS PLEASE

As opposed to graphic equalizer faders that just correlate a fixed center frequency and a gain value, parametric equalization follows its name by providing all the bandpass filter parameters. Stand alone analog parametric equalizers provide a few filter sections per channel, and may even include shelving filters at the high and low frequency extremes. Each parametric filter section should include adjustments for gain, center frequency and Q. Most users are accustomed to about +/- 15 dB of gain adjustment, and the newer digital equalizers portray a beautiful frequency versus gain display to show nice slopes as you slew gain for peaks or dips. Most users will mainly insert dips, and maybe use a few judicious broad peaks to counter a microphone or speaker deficiency in audio frequency response.

Center frequency adjustment is more than dialing up the most offending frequency and tweaking gain and Q controls. Sometimes nearby filters may have to be "biased" higher or lower to get the desired level in between slopes. In this case, having the digital display, or a high resolution Real-Time Analyzer (RTA) and pink noise source will be justified. A good example of this biasing is setting a constant directivity response along with a horn resonance notch. Typically, a high frequency horn driver will need a 2 to 4 kHz dip to cancel out a diaphragm mass resonance. But at the same time a broad (+6 dB/octave) upslope is needed after 3 to 4 kHz to handle constant directivity horn losses in the 4 to 12 kHz region and driver efficiency losses in the 8 to 16 kHz region. Passive speaker crossover networks attempt to do the same thing, but a well-crafted set of parametric filters works much better. But getting the resonant dip and the upslope blended in the presence band takes some practice.

GIMME A "Q"

The most difficult control to understand on parametric equalizers is the "Q" or Quality Factor control. Also known as bandwidth control, the Q control shapes the steepness of the filter slopes along with the gain control. For a given center frequency (f_c), Q is described as the opposite of bandwidth (BW), with Q rising as the bandwidth lowers. This is described in the equation $Q = \frac{fc}{bw}$. Since bandwidth and center frequency are valued in Hertz (Hz), Q is a unit-less quantity. Also, this filter Q should not be confused with the Q variable used as a directivity factor in speaker patterns.

Typically, filter Qs run from tenths to tens in value when used for audio frequency work. For example, a Q of 5 applied to a center frequency of 1 kHz yields upper and lower frequencies of 900Hz and 1.1 kHz respectively. This 200 Hz Bandwidth is described as the difference in half-power (-3 dB) frequencies of the filter.

Where Q shines and bandwidth fails is when the filter gains get lower than +/-6 dB. This is because the -3 dB points of the filter corner frequencies (=2^{N}and f_{low}) start melting into the unity gain baseline. For example, a filter at 1 kHz with bandwidth of 500 Hz (Q ~ 2) and again of +1 dB has no half power points, as the filter stops at the unity baseline long before the 790 Hz and 1290 Hz points for -3 dB response.

When setting Q controls on parametric equalizers, you will likely dwell in the 0.5 to 3 range. Most Q values above 5 get "ringy" when used as gain boosts, and begin to sound non-musical as the filter stored energy does not dampen down quickly when excited by a harmonic in the filter passband. However, high-Q notches are popular for nulling hums, buzzes and feed back frequencies. Very low-Q boosts or notches (Q < 0.5) are not typically used since they effect wide portions of the total audio spectrum, and are best addressed in the overall gain structure as an all-band boost or cut. In this case, reverse logic should prevail and a sound person should attack the low and/or high bands instead of a low-Q mid-band boost or cut.

OCTAVES

Now that I have got your head spinning with Qs, bandwidths and frequency corners it time to introduce the concept of octaves. While electrical engineers designing filters love Qs and bandwidths, musicians love octaves, as they relate to even-order harmonics that are pleasing to listen to. But even most of us can relate to a 400 Hz to 800 Hz "octave," because of the doubling and halving thought process. However, converting from Q and bandwidth, which are numeric base-ten processes, to base-two octaves can be reason for reaching for a calculator. This is because bandwidth to octaves conversion is in the form of equations like $f_{high}/f_{low} = 2^N$ where N is the number octaves between the f_{low} and f_{high} corner frequencies.

However, another reason to adopt octaves over Q is that DSP designers love base-two numbers, since binary represented numbers (ones and zeros) can multiply by two (one octave up) when shifted up a binary bit, and divide by two (one octave down) when shifted down a binary bit. These multiplies and divides in binary are the cornerstone of DSP functionality, and that is why you see octaves readily displayed on processor screens. I have graciously provided a handy reference chart **Figure 20-1** to help navigate from octaves to Q to bandwidths with the convenient 1 kHz points included for scaling to other frequencies. I know by publishing this chart I will piss off more than a few system engineers for revealing the relationships (keys to the kingdom), but sharing the knowledge helps us all in the long run.

The reason I described the above information is to avoid brain-lock, especially when you confront a value that you do not normally use. The best idea is to frequently review Q, Octaves and Bandwidths, so that transferring processor settings from one unit to another does not have to be a difficult chore.

| Octaves | Q | Bandwidth (Hz) | @1000 Hz | |
			F_{low} (Hz)	F_{high} (Hz)
0.05	28.85	35	983	1017
0.1	14.42	69	966	1035
0.2	7.21	139	933	1072
0.3	4.8	208	901	1110
0.4	3.5	286	871	1149
0.5	2.87	348	841	1189
0.6	2.39	418	812	1231
0.7	2.04	490	785	1275
0.8	1.78	562	758	1320
0.9	1.58	633	732	1366
1	1.41	709	707	1414
1.1	1.28	781	683	1464
1.2	1.17	855	660	1516
1.3	1.07	935	637	1569
1.4	0.99	1010	616	1625
1.5	0.91	1099	595	1682
1.6	0.86	1163	574	1741
1.7	0.8	1250	555	1803
1.8	0.75	1333	536	1866
1.9	0.71	1408	518	1932
2	0.67	1493	500	2000
2.5	0.51	1961	420	2378
3	0.40	2475	354	2828
3.5	0.33	3066	297	3364
4	0.27	3750	250	4000
4.5	0.22	4547	210	4757
5	0.18	5480	177	5657

Figure 20-1

21 FILTERS IN SPEAKER PROCESSORS

June 2005

These days, when confronted by a digital speaker processor or digital crossover you have multiple choices in the High Pass, Low Pass and Crossover Filter selections. And for those not up on filter lingo, words like Butterworth, Chebychev, Bessel, Elliptical and Linkwitz-Riley sound more like a European law firm than filter types. So for those of you not possessing an Electrical Engineering degree major with Control Systems minor, this chapter will introduce you to introduce basic "pass" filters, and help you make some basic choices in setting up a drive processor.

PASS/CUT FILTERS

The most confusing part in filters for processors is the highs and lows when combining those terms with pass and cut. In most cases High Pass and Low Cut mean the same filter. Also, the High Cut and Low Pass terms mean the same. The confusion is like the glass half full/half empty quandary. From a 20 Hz perspective, High Pass seams to have more meaning to a filter at 40 Hz. But at 100 Hz, Low Cut seams to have better meaning to the same 40 Hz filter. Eventually, the terms should merge in your head if used often enough.

Then there is the Crossover Filter lingo, meaning two filters of opposite types for splitting the audio band for peculiar types of speaker cabinets. Taking a typical example, a subwoofer cabinet may be set up with a High Pass filter (Low Cut) for about 40 Hz that matches up with sub cabinet's capabilities. Then a High Cut filter (Low Pass) at 100 Hz becomes half of a 100 Hz crossover point so that the top box cabinet starts at 100 Hz. Thus the top box filters have a High Pass at 100 Hz, and if in passive mode of operation, a High Cut at around 16 kHz to keep the dogs from barking. Crossover filters generally are set at the same frequency to avoid gaps in coverage of the audio frequency spectrum. But if the efficiencies are radically different of the adjoining frequency drivers, there may be a modest gap in the crossover frequency filters for best matching. This is usually found in two-way bi-amp top boxes with cone and compression driver crossover points and implemented in a passive crossover network.

In filter lingo, the filter name is typically associated with a number for either the number of filter "poles" or the attenuation steepness. In the bad old analog days, filter poles typically corresponded with the number of inductors and capacitors used in a passive filter circuit. Each pole contributes a 6 dB drop per octave from the filter frequency. So a two-pole High Cut filter at 1 kHz would be down 12 dB at twice its frequency, or 2 kHz. So with the more filter poles, the steeper the filter drop-off, in general. But adding poles or filter steepness does not always benefit your sound. Beyond two to four poles, some filters start ringing and creating electronic induced overtones that are usually not musical (or desired). Filter ringing is great for keyboard synthesizers, but not for sound systems.

Most typical speaker processors offer choices like: Butterworth 12, 18 and 24; Bessel 12, 18, 24; and Linkwitz-Riley 24. These captions indicate the filter type name and the dB/Octave filter steepness in the "stop-band". I tend to fall into convention and choose Linkwitz-Riley filters at the crossover frequencies, and select 12 dB/Octave Butterworth filters for the Low Cut and High Cut points (e.g., 45 Hz and 16 kHz respectively). One last tidbit to remember generic filter types is that the frequency associated with the filter is typically at its 3 dB point. For our general discussion of stop filters for speaker processing, the -3 dB point usually denotes the filter's corner frequency.

BUTTERWORTH

No, it's not a brand of pancake syrup, but a "maximally flat amplitude" type of filter. Doing my best to avoid confusing engineer terms, Butterworth filters do their best to keep a reasonably sharp but smooth drop at the filter corner frequency, at the expense of letting the phase response wander a bit. Most filters are characterized by their behavior when exposed to impulses of signal (spikes), transient steps in signal (step response) and how much phase delay the filter imposes on signals in various parts of the passband and stopband (group delay). Butterworth filters in the two-to four-pole variety has slightly more phase change than Bessel filters, and more prone to ring (dampen less) with impulses and transient steps.

Application-wise, Butterworth filters are pretty much the default choice of the far ends of the audio spectrum to ensure out-of-band signals roll quickly off. Because the phase shifts are more dramatic, you are less likely to use Butterworth filters at crossover points. There are exceptions, as you will see in the Linkwitz-Riley description, but if you want a 90 degree shift at the corner frequency, a two-pole Butterworth filter is your ticket.

BESSEL

Bessel filters are the natural opposite of Butterworth filters, in that Bessel filters are described as "maximally flat delay." This means the amount of phase shift in the passband to stopband is minimized compared to other filters. Also, Bessel filters have more dampening (less ring) when exposed to impulses and step transients. Compared to a Butterworth, a Bessel filter has a noticeably less sharp of a corner in transition from passband to stopband.

Bessel filters are best used in applications where crossover points require minimal phase change from one driver to another. But Bessel filters are the least used, thanks to the introduction of the Linkwitz-Riley filter in about 1976. Bessel filters are available, but used much less these days. If you suspect you have too much ringing in your Butterworth filters, switch to Bessel filters to stomp out that problem.

LINKWITZ-RILEY

Mr. Linkwitz and Mr. Riley are two electrical engineers that worked for Hewlett-Packard back when it was a dominant test equipment manufacturer and not the printer and computer maker it is today. The Linkwitz-Riley filter concept came from cascading pairs of identical two-pole Butterworth filters in Low Pass and High Pass configurations. The result is a crossover filter that uses the same corner frequency in High Pass and Low Pass, has no peaks or dips and is phase continuous at the crossover frequency. In other words, both drivers are in phase and each contributing half the energy they would in their passbands at the crossover frequency.

Before Linkwitz-Riley filters, the Butterworth and Bessel filters would result in differing phase motions of the drivers and peaking in amplitude response as both drivers contributed more fully at the crossover frequency. Since the birth of Linkwitz-Riley filters and modern analog "active" filters, most all analog crossover units use the 24 dB/octave Linkwitz-Riley filters at each crossover point. With the new DSP speaker processors, a bunch of analog circuits and quad-matched potentiometers are now replaced by a digital math equation that creates all these filter types with perfection never dreamed of 30 years ago.

FINAL JOTTINGS

I have included **Figure 21-1**, which shows the amplitude response for each the three popular filter types used in DSP speaker processors. These graphed response curves centered at 1 kHz are a visual example of what has been mentioned above. Whether it is line arrays or conventional groupings of speakers, the Butterworth, Bessel, and Linkwitz-Riley filters should be able to fulfill your needs. Sharper filters can be used, but carry the risk of ringing or running out of DSP capability. However, sharper filtering may prevent driver overheating and over-excursion if pushed to the system limits.

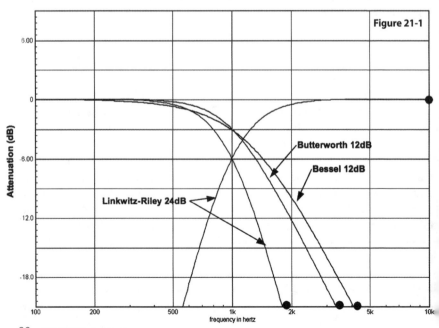

My favorite saying is, "Everything is an antenna, just some things better than others."When it comes to wireless microphones for live sound, the rules of the game are stacked against wireless mic manufacturers. The best we can do as users and operators of wireless mics is to understand the situation, and game the system to best advantage. This chapter takes a practical look at how best to operate wireless mics and introduces you to some of the rationale into how things are done.

LIMITATIONS

When it comes to handing out wireless frequency spectrum, the Federal Communications Commission (FCC) goes where money and lobbyists go. So performers, production companies and wireless mic manufacturers have little chance competing for Radio Frequency (RF) space against network broadcasters and government/military needs. What is left is permission to operate "unlicensed" on top of the UHF television spectrum at very low transmit powers and hopefully in areas where nearby broadcasters are not using their licensed spectrum. And even that is in danger of going away now, as the consumer electronic industry is lobbying to take it over when TV switches to all-digital transmission.

Until that happens, though, you'll still have to make do with the already sparse few tens of milliwatts of power coming out your wireless mics, which means only tens to a couple hundred feet of usable transmit distance. And we are not talking about the citizens band days where you might get away with "modding" your system get extra range. We have to stay within the rules, but there are things you can do to ensure the least amount of dropouts.

LOCATION, LOCATION, LOCATION

The biggest quandary is where to locate the wireless mic receivers, at monitor beach or at front of house. While FOH generally has clear air between the performers and the receiver antennae, monitor beach with its side of stage location is usually the best for wireless mic receivers, because distance between the mic transmitter and receiver is everything. But if there is no obvious clear air path between the transmitter and receiver, an alternate location should be found.

The reason for this is that performers often gyrate around, and putting the flesh of the performer between the mic and the receiver can drop an ultra-high frequency (UHF) mic transmit signal greatly. If they do this with the receivers nearly onstage, it is possible to still maintain the communication link even with the mic signals having to travel through their head and upper torso. I think of this every time I see a performer twirling around and singing through a wireless mic. Doing this twirling routine with 75 feet or more to the mic receivers at the FOH position is almost a guarantee for link dropouts.

Two phenomena account for this. The most important is the nearly spherical, or omni-directional, wireless mic antenna that spreads its 50 milliwatt signal (+17 dBm) in all directions, which means your receiver traps a very small portion of the total transmitted power. This antenna needs to have a wide spread of power, as the performer can't be asked to constantly aim the mic or its antenna at wherever the receiver is; but it also means that the expanding signal loses power at an exponential rate with distance, and that is why distance trumps all other concerns.

The other phenomenom that impacts the distance rule is the attenuation losses of RF signals going through the human body. While the human body is not equally absorbent in every area, we are mostly made of salt water (saline), and the saline absorption losses are about -55 dB per wavelength in the mid-UHF bands. Most modern wireless mics transmit in the 500 to 800 MHz bands, and make a wavelength about 15 inches in length. Since a nominal human being is about half that thickness, you can expect about a 30 dB drop in power with the receiver behind the performer. And a twirling performer could have as much as 60 dB drop if the receiver is mounted near the stage floor, since the signal has to penetrate across the torso of the performer as they face away from the receiver.

LINK BUDGETS

So far I have danced around the subject of a link budget, but as an educated buyer of wireless gear, it should be on your mind. A link budget is all the aspects of getting a signal into the RF format, out the antenna, through the intended signal path, to the receiver antenna and converted from RF back to the intended audio signal. I mentioned the first part of the budget, and that is the maximum +17 dBm (50 mW) transmitted Frequency Modulated (FM) signal out the mic transmit antenna. Now, there is some link budget just getting out of the antenna, but I will let the manufacturers worry about the FCC maximum transmit power in unlicensed operation.

On the receiver side, each antenna and receiver has a published minimum detectable RF signal strength (or sensitivity) for normal audio performance. This also entails several internal antenna and receiver link budget items, but the combination is a fixed decibel number referenced to a milliwatt (dBm). Most wireless mic receivers have minimum sensitivities in the -90 dBm to -160 dBm range. Obviously, the cheaper brands will have sensitivities in the -90 dBm to -120 dBm range, which means they need stronger signals and shorter transmit distances before dropout. So look at the receiver's minimum sensitivity number if longer distance performance is required.

The remaining items in the link budget are controllable by us. We already discussed distance, but not quantified values. If you went cheap and got a -90 dBm sensitivity receiver with the +10 dBm mic transmitter, you have a 100 dB link budget before dropouts. If everything else is perfect, a 20-foot mic to receiver separation is 27 dB of loss, and 80 feet is 39 dB. But the remaining 60 dB or more of link margin can be eaten by more than twirling performers. Things like antenna polarization come into play.

POLARIZATION

Antenna polarization is one the key reasons why diversity receivers exist. The best way to explain polarization is to imagine your typical FM radio tower and antenna. Most such antennas are just vertical pieces of metal sticking high into the air. This kind of antenna is defined as vertically polarized, since the electro-magnetic waves emit from a vertical wire antenna. Your next homework is to take your portable FM radio and try to receive far away stations with the radio's telescoping antenna vertical and horizontal. What you will find is that a vertical orientation is optimum and more horizontal orientations achieve less quality of reception.

The same applies to wireless mics and their receivers. **Figure 22-1** depicts diversity wireless mic receivers with their dual antennas oriented for best polari-ties. Without a diversity receiver (one antenna), the single receiver could have dropouts if the wireless mic antenna is perfectly orthogonal to the reciever's. Now, most wireless mics never get into this kind of trouble, but if the antennae orientations are of opposite orientations (orthogonal), the signal may have tens of dB signal strength reduction, if not a complete non-reception.

So diversity receivers with their antennas mostly orthogonal, as shown in **Figure 22-1**, will handle any mic antenna polarization from any performer. And since it will be impossible to confine the performer's mic positions, diversity re-ception will have every polarization covered (horizontal and vertical), and that means that no one receiver will get an RF polarization loss beyond 45 degrees to avoid deep signal attenuations. Also in the figure, the incorrect antenna con-figuration only benefits the redundancy of receivers if they get unequal signals of the same polarization, which is of much less benefit than if the antennas are tilted a bit.

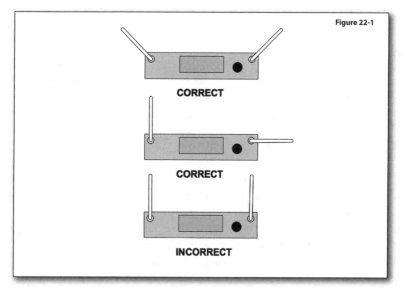

Figure 22-1

CORRECT

CORRECT

INCORRECT

In Chapter 22, I described wireless mic limitations, preferred locations of mic and receiver, RF link budgets and antenna polarization. In this part, I want to cover radio frequency names, antenna patterns and compression/expansion circuits.

SUPERLATIVES

Us techno-geeks are quick to toss out acronyms like VHF and UHF without a care, and hope that no one will bother to ask why a wireless mic might choose a VHF or UHF frequency band for transmission. Starting with HF (high frequency), we move up to VHF (very high frequency), then on to UHF (ultra high frequency), then to SHF (super high frequency), and believe it or not, EHF (extremely high frequency). Each of these high frequency superlatives stand for a range of frequencies, or bandwidth. HF has 3 megahertz to 30 megahertz, VHF has 30 megahertz to 300 megahertz UHF has 300 megahertz to 3 gigahertz, SHF has 3 gigahertz to 30 gigahertz, and EHF has 30 gigahertz to 300 gigahertz. And these are all waves per send of radiated electro-magnetic energy, better known as radio frequencies (RF).

I am blathering about Hertzian waves and superlatives because these waves travel from transmit antennas to receiver antennas at 300,000,000 meters per second. So, if your transmit frequency is 300 MHz, then a new wave starts just as the previous wave expands to one meter beyond the antenna. Much like sound waves, one meter is the "wave-length."

ANTENNA THEORY

Most wireless mics and wireless mic receivers use wire antennas cut to a quarter-wavelength of their nominal transmit frequency bandwidth. That is why VHF antennas tend to be the telescoping whip types at the receiver and floppy wires at the mic or belt pack. UHF antennas are much shorter and tend to be non-telescoping rods on both the wireless mic and receiver. The reason for only a quarter of their wavelength in size is that they use the groundplane to bounce and balance the RF waves for efficient transmission and reception. See **Figures 23-1** and **23-2** for block chart depictions of a UHF receiver and wireless mic.

The groundplane for the mic receiver is the chassis and any other conductive rack-mounting materials. The metal mic grip is the ground plane on most wireless mics. With quarter-wave antennas and quarter-waves or more of nearby groundplane RF waves can be transmitted and received well. But the mechanical configuration of the wire/rod/whip antenna and the groundplane defines the way most of the RF energy travels to/from the antenna. This shaded area in **Figures 23-1** and **23-2** is the antenna pattern of best transmission and reception, much like mic and speaker patterns. Note the steep nulls at the ends of the wire antennas where not much energy is transmitted or received. Knowing the nulls is what I want you to take away from this artwork, so you know not to locate a receiver/transmitter RF path where these nulls are occurring.

This means the receivers should be not high above the stage, or well below the stage. Also, do not place the receivers down the "boresight" line of the typical mic handling angle (down-front, behind-above). But that still leaves a lot of front/back/side locations to place your receivers for good reception. As I mentioned in Chapter 22, you typically have little to no control of how the artist handles the wireless mic, but you can locate the receivers reasonably close and out of the null paths.

THE OL' SQUEEZE-OLA

To get from the microphone diaphragm to the console, wired mics have — at most — a transformer and wire in the path. But most wireless mics have amplifiers, VHF or UHF transmitters and receivers with frequency modulation (FM) to keep the static out in their own signal chain. Furthermore, FM transmission through the air only has about 50 dB signal to noise ratio, and most wireless mics require about 90 dB or better signal to noise ratio to be at parity with wired mics. To get the extra tens of decibels of dynamic range, a compressor circuit in the wireless mic circuitry squeezes the 90 dB down into the 50 dB FM channel. At the receiver the reverse must be done by expanding the 50 dB back to 90 dB with an expander circuit. **Figure 23-3** shows a typical block diagram of wireless mic and receiver signal paths.

While compression and expansion circuits are well known in audio gear, making a very high quality compressor integrated circuit that will fit inside a small microphone is quite a challenge. Then factor in problems like very small battery voltages and the compressor must lightly "sip" power from the same battery to not drastically shorten the talk time on the wireless mic. This is why most wireless mic manufacturers spend precious engineering resources making their compressors and expanders the best — and precious marketing resources making it known that they are, and only making comparisons to wired mics nearly the same as theirs. So note those bandwidths, distortions, dynamic range specifications and battery life when comparing wireless mic systems.

VHF VERSUS UHF

Many years ago VHF wireless mics were the only affordable option. But as the RF electronic technology improved, UHF mics came about. UHF wireless systems offered smaller antennas and a less-crowded frequency spectrum to be exploited. But as I mentioned in Chapter 22, higher frequencies like UHF undergo more attenuation when penetrating performers in the path of the receiver. For awhile, this made VHF wireless less expensive and more reliable, until UHF receiver technology could handle the weaker UHF signals. Today only a few VHF wireless mic offerings are still available, and most are for cost conscious music industry (MI) grade products that can live with the bigger antennas.

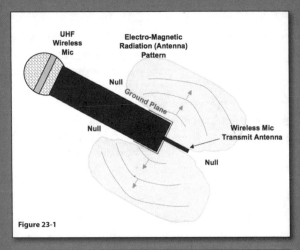

Figure 23-1

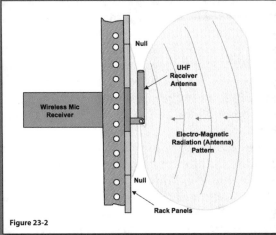

Figure 23-2

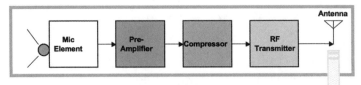

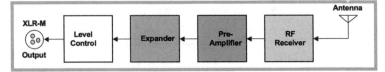

Figure 23-3

PRACTICE

The
practice chapters
are the catch-all chapters
that did not fit in the other
sections, and mostly deal with the
execution aspects of doing a good
show. Having great gear does not
ensure a great show, but a great sound
person can make poor sound gear
perform to the best of its abilities.
Hopefully, you will find tidbits
of handy tips to make your
practice all that much
better.

Acoustic reinforcement is our primary task as sound system operators. To the layman, though, we are perceived as the dispensers of loudness who make the performance memorable and worth the ticket, as opposed to background noise. With all the new technology available to us, I believe we sometimes get lost in technical abstraction to the detriment of the ticket buyers.

To illustrate this, let's go back a half century and peek at the loudness chart that physicists Fletcher and Munson created **Figure 24-1**. In this chart the curved lines running horizontally across the chart indicate the dB level that frequencies need to played at in order to be perceived at a constant volume by the human ear. For example, most of us mix for Sound Pressure Levels (SPLs) in the 80 to 110 dB region. Note that at these levels, the human ear response is pretty flat up past 1 kHz, but needs a little less 2 kHz to 5 kHz, and some help beyond 10 kHz. But if you are buying loudspeaker digital signal processors (DSPs) these days, you are given the lecture that "flat is good," and that getting a flat response across all your frequencies in terms of dBs should be the goal of your SPL measurement gear — even though that's not how the ear hears.

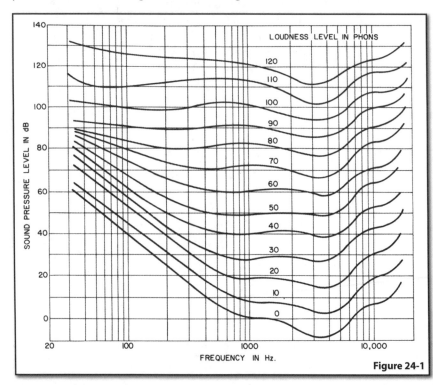

Figure 24-1

If you take this to its logical extreme, everyone should be complaining of excessive presence and not enough sizzle in the highs.

To cite an anecdote, a television listener was complaining of how awful a rock concert sounded coming out the TV in their living room. That complaint was telephoned to a soundman-friend who was watching the same show at home, and the response was "Turn it up loud!" Sure enough, the concert was a direct feed from FOH where the band engineer was bathed in +100 dB SPL. This instance is rare now that televised concerts usually get a feed to a remote truck with an audio engineer mixing at living room levels. But it goes to show you that loudness is a perception game.

The point to take with you is that the gear can be optimized to be flat response, time aligned and capable to delivering great volume. But as a FOH engineer, you are responsible for taking the mix to those ticket holders and make it sound great, not just pretty on the metering.

In the February of 2003 issue of *FRONT of HOUSE* Paul Freudenberg wrote about the inverse square law effects on measured SPL, and OSHA's workplace limits on SPL durations.

The challenge is to apply this knowledge while mixing a show. For example, my usual club FOH position is about halfway in the room with, the stage/dance floor at one end. Typically, the closest seating to the FOH speakers are two "halvings" towards the stage from the FOH position, or about +12 dB SPL more than I measure at the console. This also means that these seated patrons hear more than twice the loudness I do during the show. So the nagging question is, "Could I stand the loudness if I were seated in front?"

That is why metering SPL is a good technical guide to remind you of the nagging question. Of course, OSHA is not likely to shut down the performance, but they will give you and the venue employees a stern lecture on day-after-day exposure, and why wearing ear plugs is a good habit to start. In monitoring SPL, choose the OSHA "C" weighted scale as it best fits its purpose at levels of 85dB and beyond. C-weighting is flat from 80 Hz to 3 kHz to match up with the Fletcher-Munson curves at louder levels. The "A" weighted scale is intended for quiet readings typically in the 0 to 55 dB range, and has the 300 Hz and down information filtered out to match the Fletcher-Munson curves at low SPLs. There is a "B" weighted scale to handle the intermediate SPLs, but it is rarely used.

A-weighted SPL meters are more likely to be used by law enforcement officials when checking into disturbance calls. Many communities have evening restrictions on nuisance levels, such as no more than 60 dB SPL (A-weighted) from a 100-foot distance of the offending source (building, dwelling). This also should be a factor in your loudness judgments inside of thinly walled clubs or outdoor gigs.

THE NEIGHBORS, THEIR LAWYERS

Our modern, litigious society is no longer laissez-faire about loudness and how the youth are ruining their hearing. "So sue me, baby," is no longer a light-hearted quip, but a reality in the live entertainment business. We have tobacco lawsuits, fast-food lawsuits and soon-to-be concert lawsuits. I find it less than amusing that lawyers are trying to curb the true entity that the producers are marketing according to what the consumer desires. Whether it is soothing cigarettes, tasty burgers or a loud, exciting concert, the whole notion that personal responsibility and moderation are now exempted offends me.

Let me cite a couple of related examples. *FRONT of HOUSE* magazine's editor, Bill Evans, used to live in Pasadena, California, a couple of miles from the Rose Bowl. Sure, in the 1920s when it was built, the nearby residential neighbors could expect the Saturday afternoon roar of a football crowd and a marching band and that was about it; but do today's Rose Bowl neighbors have the right to complain if a rock concert is held in the same venue under reasonable SPL and time-of-day restrictions? I say, you bought the house in this neighborhood, you choose your fate — and live with No Doubt cranking it out at 50 dB SPL on your lawn at 10 p.m. Now, 50 dB SPL is half the loudness of a quiet conversation, but if you do not appreciate the music genre and like to sleep with the windows open, you have a problem.

The other example was a debate in my area about a proposed amphitheatre being built on a former garbage landfill and facing into a Mississippi river wildlife refuge. The landfill wouldn't have been suitable for other residential development nearby, and opponents claimed that the noise would disturb the habitat of the refuge. I say this is a perfect amphitheatre spot: There's no human neighbors, and the river bottom wildlife who are present will be disturbed less than 1% of the time because the seasonal use and the few shows per year. If the deer population can't handle an occasional Ted Nugent concert, I know that Ted has a very pointed solution to the problem that wouldn't entail activists engaging the services of lawyers. I also find the location ironic, as it is just a couple miles from the Bosch/Telex plant (EV, Midas, Klark Teknik), which is equipped with a two-story anechoic test chamber so quiet that you can hear the blood flowing near your cochleas.

So beware of the gaze of the law profession, since they see loudness as a vice to be controlled. Like the other precedents of tobacco and fast-food, the "low hanging fruit" will be first to be sued. This will likely be event promoters, venues, sound companies and speaker manufacturers. A couple of future scenarios can be envisioned. One is that FOH console operators will be forced to sign a SPL responsibility form indicating that you shall not exceed a reference level or you are solely at fault and liable to loudness lawsuits. Another scenario could be DSP SPL monitoring at multiple locations and some kind limiter/governor circuit taking over control of the mix loudness from the console. The industry has spent nearly a century developing ever more efficient loudspeakers, now those same manufacturers have to shore up their legal defense funds, and teach us users how to keep our patrons safe from hearing loss.

25 | MADNESS TO THE METHOD, GAIN STRUCTURE

June 2003

In this dissertation I am going to do a little Q&A, á la the old *Audio Cyclopedia*, about questions you may have wondered about while you just did what you were told. I am going to throw down a generous helping of electronics history as a way of answering the questions, and to remind us all were all this technology came from.

Question 1: *Why is 0 dBu the reference signal level?*

This question could easily be re-phrased "Why is 0.773 volts RMS (0 dBu) the standard and not some other convenient number like 0.1, 1 or 10 volts?"

The answer goes way back to Alexander Graham Bell's era, when no such thing as radio or broadcasting was thought of. As the "Bell System" and "American Telephone and Telegraph" (AT&T) became the monopoly in the phone service industry, Western Electric Company was formed as a subsidiary of the Bell System to design and produce telephone gear for the whole country. After much trial and error, a standard two-wire pair transmission line with 600-ohm source and load impedances to maximally send carbon microphone signals down the wires was developed. With the right construction materials, voice signals (about –20 dBu) could transit 5 miles with a passable loss of signal amplitude.

When Lee DeForest invented the Vacuum Tube Triode for signal amplification in 1906, his killer app was boosting feeble telephone signals, thus creating long-distance phone service in the second decade of the last century. Western Electric still had a lock on the electronics industry in the 1920s as broadcast radio was just emerging, so naturally it had the highest technology suitable to fulfill civilian and military requests for standard "Public Address" apparatus. By the early 1930s Western Electric had the first quality dynamic microphone (requiring no DC power, unlike carbon mics) and combined vacuum tube amplification connected to the first efficient "loud-speaking apparatus" that we now know as horn loaded drivers.

As broadcast radio became widespread, and specialized companies like Electro-Voice, MagnaVox, and Shure Brothers came to supply (with Western Electric) the needs of public address and broadcast gear, the 600-ohm line cabling still held as the lowest loss method of distributing and processing audio signals. From that era, a one-milliwatt reference level into 600 ohms became the reference level, or 0 dBm (zero deci-Bels referenced to one milliwatt). When you measure it, 0 dBm is exactly 0.773 volts RMS. But as technology marched on, audio electronics moved from power matching to "bridging" impedance matching, and the 0.773 volts without any specified load impedance was now described as 0 dBu (zero deci-Bels unreferenced).

To answer the lingering question of what became of Western Electric, government anti-monopoly policies in the 1930s forced the breakup of AT&T (the first time) into RCA for broadcast, Bell Labs for telephony, and All-Technical Products (Altec) for public address. Altec slowly became Altec-Lansing, then split back to Altec and James B. Lansing Inc., then on to JBL.

Question 2: *Why gain (trim) to line level's and then mix afterwards?*

This question is more math than history, but we can still thank the early broadcast pioneers of the 1930s for the first work on defining signal-to-noise and noise source definition. This question could also be stated as: What is the best method to minimize hiss in the mixing console?

The answer comes from the invention of the radio and techniques used to maximize signal-to-noise ratio and thus transmission distance. As a signal is created, processed and sent to its final destination there is a signal-to-noise ratio (SNR) degradation. As each stage, or processing block, passes the signal, the noise eventually encroaches on the signal level. The number of dB drop of SNR per stage is defined as its noise figure (or noise factor for you dB challenged). A noise factor of 6 dB or less per amplification (gain) stage is considered a low-noise design for a preamp.

To better visualize this idea, lets put some example numbers to work. If a typical dynamic mic with human voice provides –50 dBu signal peaks, and the console's referred input noise is –128 dBu, you have a 78 dB SNR, which is respectable in live sound applications. As the signal proceeds through the channel mic preamp, EQ section, channel fader or VCA, summing amps, master fader and balanced line driver, there is a noise figure penalty to be paid. The good news is that if two gain stages are cascaded together, the noise figure of the first stage dominates, with the second gain stage noise figure effectively divided by the gain of the first. What this means is that cheaper electronics can be used after the mic preamp, with a high gain preamp covering for the sins of the rest the console's electronics.

One other item to be shared is that attenuation circuits (EQ filters, faders, pots, VCAs, etc.) can generally be assumed to be direct losses in SNR, with every dB in attenuation a corresponding dB increase in noise figure. So the theoretical perfect (low noise) mixing console setup would be faders maxed, EQ flat and amplifier gains a perfect match between mic level and power amp full power sensitivity.

But no realistic scenario exists on a mixer without faders to mix with. So the next best answer is take your desired loudest channel in the mix, set its preamp gain control (gain, trim, etc.) for about 0 dBu average level, and keep the channel, group, master faders reasonably high, but preserve some headroom for the occasional "louder" demand. This minor compromise yields the best SNR while still giving mix flexibility. This practice also applies to the gain of all the other signal source channels, but with the obvious idea that their faders would be more attenuated.

Question 3: *Why is +22 dBu the common maximum level?*

This answer also comes from electronics history, but only a half-century back. The dawn of the first mass-produced transistors had a typical maximum voltage level of 30 to 40 volts. Of these early transistors, many were targeted for industrial controls and analog computers for military and aerospace usage. The most common analog computer section was the operational amplifier or "op-amp." Because these op-amp sections were designed to be near perfect mathematical gain stages, they had both positive and negative voltage swing capabilities. By taking the limitations of the transistors plus the need for a bi-polar (plus and minus) power supply, the standard of +/-15 volt supply levels was instituted, and is still used today.

As transistors got grouped on one silicon die, integrated circuits (ICs) were born, with the first standard products becoming IC op-amps. As IC prices dropped in the late 1960s and early 1970s, more IC op-amps started finding their way into audio equipment, still requiring their +/-15 volt power supplies. Today's pro-audio signal processing and mixing gear is largely composed of IC op-amps and a few application-specific ICs, plus just a few necessary un-integrated transistors. The common legacy of supplying them with +/-15 volt levels still exists, with op-amps capable of near +/-14 volt audio signal swings. This level translates to about 10 volts RMS, or +22 dBu at which the circuits would exhibit clipping of the signals. Some math trickery may be also in maximum output specifications as you can gain another 6 dB in level by stating the output as balanced, in which each balanced output contact swings in opposite polarity to double the levels.

With the above explanations that we should set our levels close to 0 dBu and keep away from the clip levels around +22 dBu, there leaves plenty of headroom for classifying what peak signals can be and what is required to get the drive channels (EQs, crossovers and power amps) to full output. Most power amp manufacturers set their sensitivity values to around 0 to +9 dBu for full unclipped speaker drive. Keeping the post pre-amp levels at or below the power amp sensitivity values mostly assures a clipping free production. Of course, that relies on keeping the power amp input attenuator controls full up. You do not want to be low-noise all the way to the amp, and then throw away all that SNR at the last attenuator do you?

FINAL WORDS

Our latest generation of audio production personnel deserve to be educated on how we use our gain structure procedures, and why these methods came about. Some may argue about exact levels and forming up mixes, but I am coming at this from an electrical engineering viewpoint, and am attempting to shine light on what the design engineers consider optimum use, rather than "that seem to work for me" operator tactics. We need to appreciate that live sound borrowed heavily from the telephone and radio broadcast pioneering work, and that electronics achievements impacted our practices. The 20th century could be termed the "electronics" century, and it looks like the 21st century will be the "photonics" century — with fiber-optics promising near unlimited bandwidth for passing analog or digital signals from baseband to radio-frequency to optical-frequency signals.

When chapter 25 was first printed in *FRONT of HOUSE*, I hit a nerve with a few readers' philosophies of drive levels exiting the mixing console and entering the power amplifiers. Those who commented on my idea of keeping drive levels near the 0 dBu mark rebutted using the rationale of Signal to Noise Ratio (SNR) optimization and the idea that every piece of gear in the drive line should reach clipping at the same point.

For me, this puts us into two camps. I belong to the camp of old codgers who lived and died by the VU meters, with everything defined as line level never pegging the needle to the end stop (about +4 dBu). Then there is the "run it hot" crowd, who believe it ain't clipped until every last red LED on the LED ladder has been lit. Both practices are valid, but each has their downsides operationally. Obviously, the run hot crowd has to back off the amp gain controls by about 10 dB or more. The ol' codgers can leave the gains at full tilt, and never have to worry about someone getting the "crank it up" impulse at the amp rack.

RUNNING THE NUMBERS

To partially justify my bias towards keeping my levels cool while throwing a few dB of SNR away, I look at the highest line level I need to achieve to get every drop of power from my power amplifiers. Right now, I have been using a rotating group of subwoofer amps set up at 32 dB of gain (40V/V) that are the QSC PL-6.0 II and the Lab.gruppen fp6400. Both are capable of delivering about 2200 watts into a 4-ohm load, which in this case happens to be a pair of 2 by 18-inch subwoofers. Taking the power (2200 w) multiplied by the nominal load (4 ohms) and performing a square-root function leaves us with 94 volts RMS output drive to the speakers. Dividing the output by the 40V/V value gives us the 2.35 volts peak input signal, or about +9.7 dBu.

At this point, the "run it hot" crowd will dial in a −10 dB setting at the subwoofer amp so that they can put +19.7 dBu at the amps' XLR jack to get +9.7 dBu at the amp's gain control wiper contact. Us cool campers are running the amps wide open, and still bending the rules by having to run into the "yellow" to get +9.7 dBu. Granted, most us are not pushing the limits in most gigs, and tend to leave the +9.7 and +19.7 dBu levels as our signal peaks, while leaving the average 6 to 12 dB lower as average levels.

NOISY FLOORS

Our run-it-hot crowd spares no rest to mention that limitations on most mixing consoles create noisy signal sources that need the extra 10 to 15 dB signal excursion. While my impulse is to strike out at your choice of crappy mics, crappy console and poor gain structure tactics, I do share the compassion that many times things do not roll our way, and noise can be a constant companion. I usually am happy if my noise floor leaving the console is about –60 dBu. That typically means I can easily hear the hiss in the mid-and-high frequency drivers at 1 meter away with no program input, but that the hiss is pretty much undetectable when quiet program material (sound check CD, speeches, etc.) is being run through the system.

CLIP THE AMPS FIRST

Call me forever poisoned by Hartley Peavey's Distortion Detection Technique (DDT) feature, but having amplifiers with soft input signal limiters has been a godsend. Such limiting circuits on most power amplifiers today means that driver-killing clipping can be replaced by softer distortion that is audibly detectable, and serves as polite warning you have hit the system limits. This is caveat-ed with the knowledge that nothing upstream in the signal path has reached clipping beforehand. This is my first rationale why I do not hang with the run-it-hot crowd, but pay a modest SNR penalty to do so.

With the ever-increasing assurance that our power amplifiers are modern and include signal limiters, it is prudent to actually set the clip levels of the loudspeaker processor, equalizer or crossover higher than the amplifier limit levels. The run-it-hot crowd has to sweat the details of how many dB's higher, but the codgers have no worry with their +22 dBu minus +9.7 dBu as headroom. I confess to rarely paying attention to drive signal path levels, and just glance at the power amp rack for proper level indications. My rule is that amps soft-clip (limit), and EQs, comps, crossovers hard-clip; and hard clipping is a driver destroying practice.

NOT FADE AWAY

Part of getting the mixing console drive levels correct is the backward impact imposed on subgroups and input channel levels. When kicking off a gig, I start with the master faders (a.k.a. the gas pedal) at the –10 dB mark, and push up or pull back as required with the lead vocal and its sub-group riding the 0 dB marks. Small rooms with little stage wash will cause me to back off many dB on the gas, and the outdoor gigs are likely to get me goosing the pedal a few dB as the show mix settles in.

Subgroup fader position philosophy can be all over the map, depending what the console operator thinks is most important. I have tended to fall into the "leave it at the 0 dB marks during the show" camp with rare exceptions. Lately, I have been begrudgingly working the subgroups more, mostly to drop instruments in the quest of keeping the vocals on top of the mix with varying stage wash and performance dynamics. But with the subgroups near zero, I still have to keep a watchful eye on metering on the subgroups to avoid running out of headroom.

The subgroup headroom issue then trickles down to the channels. I used to love working the channels hot — in the +3 to +10 dBu range — and throw away the levels in the channel fader or gas pedal. Lately, however, I have mended my ways with cooler peak values in the 0 to +3 dBu range so that my compressors and insert effects are not running on limited dynamic headroom. By running cooler at the channels, not only do I avoid summing amp clipping at the subgroups (in non-VCA desks), but it forces me to step on the gas pedal (faders) a bit more. I find these practices avoid the problem of "running out of fader" in any signal path on the console.

FEEDBACK

The above paragraphs are not meant to be just educational, but to stir thought on the mix creation process. Please feel welcome to share your insights, tips and criticism on these fader and headroom tactics. Let us also discuss what not to do — and anecdotal stories are welcome, as are letters to the editor and posts in the FOHonline.com forum pages.

27 TROUBLESHOOTING

December 2003

Of all the skills a live sound operator uses, system troubleshooting is the one skill that needs to be the sharpest. It is a touch ironic that no sound person ever wants to apply these skills, but when the need arises, only the utmost in system knowledge and plenty of hands-on experience will enable you to fix problems under the intense time constraints.

Whether it is a circuit board or a complete sound system, the principles of technical troubleshooting apply. The first aspect to grasp is the documentation of the system circuit layout. In Printed Circuit Boards (PCBs), a circuit schematic diagram or "photofact" sheet provides the ins and outs, the intersection points (nodes) and expected signal types and amplitudes at key nodes. While most experienced sound system technicians do not keep paper copy system schematics, most do commit to their brains a mental picture of the sound system and the flows of signal cabling.

DIVIDE AND CONQUER

Divide and conquer is not a battle strategy for the Roman Empire, but a basic mental tool for quickly getting to the root of the problem. Given a chain of signal processing circuits with no clue as to which one element corrupted the signal (no signal, noisy signal, distorted signal), the divide and conquer strategy says to start at the midpoint of the chain and check for proper signal conditions. If the signal is still bad (corrupted), then take the first half of the chain and probe its new mid point for signal goodness. If the midpoint signal was good, go to the new midpoint of the latter half of the chain and check for signal goodness. This process repeats until the cable or signal-processing block (mixer, processor, amplifier, etc.) is pinpointed. **Figure 27-1** shows a generic sound system block diagram showing typical signal chains.

In sound systems, not everything has a single start and end point. Because mains and monitor speakers tend to split early, creating two chains, troubles that affect both systems quickly narrow the problem down to the sources (mics, DIs, etc.), cabling and source phantom power supplies. By considering the main/monitor console(s) as the convenient "midpoint" for divide and conquer, getting good signal to the console and having the means to monitor it via metering and headphones is a crucial first step. Besides our human senses and headphones, good troubleshooting tools for your briefcase are a dynamic talkback mic and some kind of audio signal generating device. For me, I keep a Neutrik Mini-rator handy, but there are other smaller, cheaper alternatives. Later in this chapter, I will note some other handy items for troubleshooting and field repair.

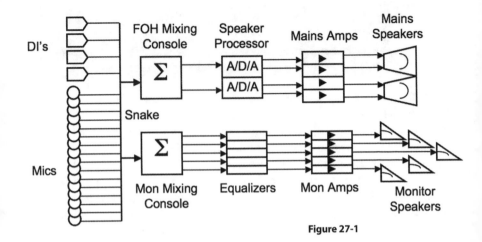

Figure 27-1

ALWAYS HAVE A "PLAN B"

As you come down to "Who Shot John?" in identifying the suspect cable or piece of gear that is malfunctioning, you should have a "Plan B" in case the item can not be put back into working order promptly. Plan B can as simple as having a spare cable or amplifier for Just-In-Case (JIC) usage. When providing a system, my JIC collection will include spare cabling of every type used, a small mix console and a small amp case with a stereo power amp plus an active crossover unit. In selecting JIC items, think of every possible item in the system and how you would manage if it failed. The response "go back and get another" should not be a possibility, nor calling for another and hoping some buddy will make a timely run for you with a replacement item.

The worst scenarios I can imagine in B- or C-rig sound systems would be the loss of a mains snake, or the mains console. If the snake is un-repairable in the field, one could abandon the FOH position and relocate to monitor beach on the side of the stage. If the main console is out and no other console exists for monitors, then a spare console or mini-mixer should be part of your JIC collection. Back when I hauled all my gear to gigs in a Ford Econoline I kept a dusty, knobless TAPCO six-channel mixer under the driver's seat for such an occasion. No sound system provider keeps a complete spare system for JIC, but a few well thought out spare items are worth the fuel expense to ride along.

COMMON TROUBLES

From my experience, here is a short list of common troubles and typical first-to-look-for fixes.

Forgot to Patch Cable(s) — As obvious as it may seem, when it comes time to "fire-up" the system and nothing comes out, it's usually because of skipping a step in patching the system. First, bang your head on something hard, then vow to stick with your routine for cabling-up next time. A variant of this is forgetting to plug things in to the power distribution system. You can claim that they are solar powered, but the stupidity is already showing.

Bad Speaker Cable Connection — The classic is the much maligned TS phone plug/jack on wedges, and the tip-sleeve short that occurs when the plug is half way out. Amplifiers show little signal output and possibly active limiter indications. Most of this goes away when using twist-lock connections.

Buzz or Hum in System — This is usually accompanied by my question, "Which of you musicians decided to plug into the wall, instead of my stringer receptacles I provided?" Getting all the signal critical cabling balanced and working on one power distribution system is the key. Having DI boxes with ground lift switches (e.g., the Whirlwind IMP-2) is handy, but when all else fails, a couple of Sescom IL-19 XLR in-line isolation transformers is the last resort.

Effects No-Show — Especially with TRS connections, getting the send and returns patched backwards at the mixer is a classic faux-pas. On insert effects or dynamic processors that mute out when inserted, remember ring is return, tip is send on insert cables, with the rare exception.

Crunchy, Distorted Sound — Once you have double-checked for proper signal amplitudes, check for dirty or corroded cable connections. Make sure you have your Deoxit or Pro-Gold cleaner handy. A couple of inserts/extracts may help rub the contacts clean as a temporary fix until the next break.

REPAIRS

If you have the luxury of time, or no other alternative, doing simple cable field repairs is a task to keep tools stashed for. Typically, a handy bag or box containing a pencil soldering iron, 60/40 rosin core solder, Suction-vise or Pana-vise, spare connectors, needle nose and diagonal pliers, shop rags and WD-40 spray is a good start. The roadie's secret weapon is WD-40 (Water Displacement, Formula 40) as it can clean some metals, as well as melt most tape residue off of cables, not too mention its automotive purposes.

When mentioning tools, other handy special purpose tools for troubleshooting include a Deoxit pen for connector contacts, a XLR-M phantom power detector (connector with a LED poking out of the boot), and a TRS phone plug or two with an intentionally shorted tip-ring link for getting past flaky TRS insert jacks.

It's that time of the year where you do your most lucrative gigs, face mother nature head-on and find out again what your rig's weakest links are. Yes, the late-spring, summer, early-fall outdoor festivals willl use all your gear — plus a lot of little things that tend to escape us until the last moment.

WHAT WAS THAT DOOHICKEY?

While most of us try to be so prepared we'll put a Boy Scout to shame, every summer there seems to accrue a memory-only list of things we said needed to be done for next year, then promptly forgot. So go back through your last year's business records and see if they bring back the "missing item" list placed in the back of your head. Forever paranoid while packing up the trailer, my motto is "there is always something that you forget." Make sure your forgotten item is something trivial. Lately, my forgotten item has been the two-wheeler handcart or "dolly." So now I am going to buy another dolly for the trailer so the other can be left back at the shop. I try to put everything on casters, but smaller speakers do not fit the criteria for casters while looking appropriate on stage.

In this outdoor season — or, as I think of it, production hell — all the world is not a stage, and thus other items are packed up JIC (Just In Case). The first seasonal JIC items to stash are wood shims and small pieces of dimensional lumber. It is pretty common for outside performance areas to be slightly sloped for rain runoff, but even a little slope or curvature can cause havoc with our nice, true-square shaped gear. So carry a couple of packs of lumber-yard pine shims to cure a case of the wobbles. The other items are a few short pieces of 2-by-4 studding for leveling speaker stacks on sloped ground. I get a little anal about it, and have the wood painted flat black since they have a habit of being in the audience site lines.

Regarding rain-wear, your gear needs some as well. My outdoor season starts with a shop rag well dampened with WD-40 to wipe down the roadcases. Not only does it remove the winter accumulated debris, it puts a modest shine on the black laminate and slows up future oxidation on the exterior case metal-work. While doing this, remember the bottom-side dolly-boards to get the insect residue and dust out of the bearings. Any loosey-goosey castors are candidates for immediate swap-out. Being a one-man crew for many calls I am extremely tough on casters, and pushing heavy casing onto ramps can bend up even 4-inch castors.

Remember all those large clear poly-vinyl bags your gear was packed in when new and out of the box? Well kick yourself real hard if you did not save the larger, thicker ones for use as speaker rain covers. We all think that it will never rain on our rig, or, if it does, that a light droplet prelude will give us the time to mount a crew rain-drill to cover everything in sight. If you ever have been caught in a sudden downpour you know better, and pre-cover everything and secure the covers for that out-of-the-nowhere gust that removes everyone's headwear. And yes, you can mix with bagged speakers, but understand you will lose some high-frequency response. If clear poly bags are not available, keep a box of industrial or lawn trash bags handy instead. Oh, and you do have a complete selection of poly-vinyl drop sheets and/or tarps as JIC, right? And they are out of their store packaging for quick rain-drill deployment, right? I thought so…

A BRIGHT, SUN-SHINY DAY

Other weather-related problems are wind and heat. Now is the time to throw a new canister of Wet-Ones in the mic locker, not only for saliva-molested mic windscreens, but for your dirty paws too. And did you remember to re-stock the road trunks with sunscreen and bug repellant? On the wind front, you did repair the holes on your E-Z Up tent last year, correct? Do you have an E-Z Up tent roof? If not, now is the time to get your order in from your equipment dealer (or Sam's Club for light duty versions). And how about remembering those 20-pound sandbags for anything prone to catch the wind? They are not just for cymbal stands.

While typically relegated to last second forget-me-nots, think about yourself and the crew while out in production-hell. Things like talcum powder (beware of Gig Butt), coolers to stow snacks and beverages, hats and footwear are extremely important for extended outdoor crew calls. Break in your summer shoes now. It may look unfashionable, but my camo jungle floppy hat minimizes head and neckline sunburn while keeping my ears uncovered. Better than those baseball cap wearing crew lobsters at the end of the day. Also, now is the time for a serious first aid kit or three in all the convenient places.

Here is a tip for all you wise production guys: Find a way to keep a small space set aside in your trunks for the "Muso Convenience Store." Things like 9-volt, AA and AAA batteries, drum keys, drum sticks, guitar strings, bass-guitar strings and guitar cables are great gig-savers when show time is imminent and trip to the store is improbable. Charging $5.00 for a single 9-volt battery is not outlandish for this situation.

SUMMERTIME MAINTENANCE

First on the priority list of summer season maintenance is your transportation. More miles get added than at any other time of the year, so invest your resources now for a worry-reduced drive time. Tire life and inflation pressures should be checked more frequently, and the spares need attention, too. Have you got that stash of essential top-off fluids, mechanic tools and shop rags? How about tire chalk for trailers?

Next, think about all that road time and the jolts, bumps and vibration your gear is exposed to. If you do not have shock-mount rack-cases, how about a resolution to start the rack swap-out process towards more shock-proof casing? Remember how you needed more racks and stacks for some of your gigs last year? Now is the time to make those capital purchases and/or call up your business-buddies to see if they can cross-rent some to you. On your existing amps: Break down the racks, remove the amp covers and carefully micro-dust the heatsinks, removing a winter's worth of indoor dust-bunnies, then check for serious fogger/hazer juice corrosion.

Before getting into too many outdoor gigs get those feeders, cords and cables wiped down with WD-40 to remove any adhesive residues. Nothing worse than dirt or sand adhered to cabling, all because you were too hasty in the spring and did not do cable maintenance. Also, all those cocktails spilt on your cables in the winter do constitute an adhesive, and insects may want to clean that residue before you do.

Remember to remember what you deemed never to be caught without last summer. The cases of bottled water and aspirin can wait until load-out, but use this diatribe as a call to action before your summer busy season sneaks up on you.

A s a sound-person, I would like to be able to walk into a venue and, using only my eyes, have a pretty good guess of what the room would sound like as far as acoustics are concerned. Most of this opinion should come from looking at the performance area, noting the stage, the floors, ceilings and walls that inevitably reflect the sound throughout the audience areas. This installment is to give you some of the tools to size up a room.

WHY?

Knowing a room's reflections is important because the performance quality is at stake, and taking actions at load-in could make for a lot less of a struggle setting levels and equalization. Using conventional speakers, the inverse square law applies, and sound waves drop at a 6 dB SPL per distance double as they emanate from the drivers. As the sound waves disperse from the speakers and stage backline, they are bound to hit the room surfaces. The ideal situation is to diffuse or absorb the waves into the surface, and minimally reflect back into the audience areas. These reflections (reverberation), if not attenuated enough with respect to the direct waves from the intended sources, tend to reduce the intelligibility or the source material (speech, music, rap, etc.).

Of course, it's preferable that some music types are reflected as part of the performance. Musical styles before the electronic age such "classical," "old time" and "polkas" use the room reflections to broaden the sound, much to the enjoyment of the patronage. Modern recorded music played back is best in "dead room" conditions, since the recording mix has inserted the desired ambiance already. So somewhere between live and dead room acoustics is where most modern music performances are done.

ABSORPTION/REFLECTION

As direct or incident sound waves hit a surface, such as a wall, some will be absorbed, some will reflect and a small amount will pass through the surface. **Table 29-1** at the end of this chapter shows a list of common surfaces in music venues and the absorption in % at three nominal frequencies of 100 Hz, 1000 Hz, and 4000 Hz. The 1000 Hz and 4000 Hz numbers are very important, as it represents the midrange and presence bands of frequencies where the human voice produces consonant sounds critical for intelligibility. A good assumption is that the majority of sound not absorbed is reflected back away from the surface.

MASONRY

A soundperson's worst acoustic enemies are masonry — concrete, brick and rock-based materials. The exception is the +3 dB boost in SPL you'll get with sub-woofer cabinets on masonry floors. Note that carpet covered concrete does great things since it absorbs mid and high frequencies well, but still leaves the subwoofer efficiency intact. Without any fabric material covering concrete, brick or stonework you have a nasty surface for walls and ceilings. Your best mitigation technique in dealing with this dilemma is medium to long-throw speaker cabinets, and aiming them away from the surfaces and downwards to the audience areas.

AUDIENCE

I intentionally placed the empty seats and seated audience absorption in adjacent rows of the chart to show the vast difference an audience makes. Since the audience is composed of fabric covered, fleshy-encased bags of saline, meat and lard — known as human beings — we are really good sonic absorbers. That is the main case to be made that pre-show sound checks show just half a picture of how the room will sound occupied. And this is also the whole reason for asymmetric horns on stacked speakers and rigged speakers aimed into the seating. The only thing people absorb better than audio is wireless radio frequencies (UHF), and thankfully no high power transmitters are allowed in venues (-52 dB/wavelength).

GOT WOOD?

Wood based surfaces are slightly nicer than masonry, but are still surfaces of high reflectivity to mid and high frequencies. One of the reasons I run around with garbage bags filled with folded sheets of black velour and commando cloth drape (flame retardant) is to cover an upstage wall of wood that creates plenty of stage wash reflections. Unfortunately, my favorite venues love knotty pine décor, and wood is part of the visual ambiance, even if the aural ambiance is less than desirable.

GLASS

Glass is another modern problem in acoustic reflections, since it absorbs little low frequency energy, and is even worse than wood in reflecting mid and high frequencies nearly perfectly. With many modern performance areas opening up to mother nature or looking like terrariums, these are more trouble spots worth avoiding. Fabric is the solution, but that is not likely because of the visual aesthetics.

BEST OF THE BEST

Nothing brings a smile on my face quicker than a dark, intimate club that tends to be a quiet setting until the performance starts. Things like acoustic suspended ceilings, lots of floor and wall carpeting, upholstered seating and no nearby side-walls at the stage are the hallmarks of these venues. Using the hand-clap test usually yields nil reflections. Big rooms of this category are live sound nirvana. If soundmen were live music club owners, they would all purchase old bowling alleys and carpet the seating areas.

Table 29-1

Material/ Surface Absorption	@ 100 Hz	@ 1000 Hz	@ 4000 Hz
Empty Steel-Padded Chairs	15%	39%	30%
Seated Audience	60%	96%	85%
Sheetrock Walls	29%	4%	9%
$^3/_8$" Plywood Paneling	28%	9%	11%
Carpet on Concrete Floor	2%	37%	65%
Heavy Velour Drape	14%	72%	65%
Wood Flooring	15%	7%	7%
Painted Concrete Block	10%	7%	8%
Thick Glass Window ($^3/_4$")	18%	3%	2%
Brick Wall	3%	4%	7%
Linoleum on Concrete Floor	2%	3%	2%

One of the true skills a sound person needs is the ability to identify audio frequencies, especially when ringing out monitors. This piece is kind of a back to basics on ringing out stage monitors, but even experienced sound techs can find this a bit of validation, as there are many ways to ring out wedges. If you have tips and tricks not mentioned here, drop us some mail and share your methods.

SQUEEEEEEEEEL......

The infamous ring that Hollywood loves to add in movies when someone approaches the podium microphone is cliché, but it does serve as an example of acoustic feedback. This comes from stage monitor or main speakers that have the sounds fed back into the suspect mics, which are then amplified again and again. Normally good speakers and good mics are the first line of defense against feedback, but most acoustic-electric transducers (speakers, mics, instrument pickups) have non-flat frequency responses that are undesired — unless the situation makes them desired.

While you may choose mains speakers for a special frequency response and coverage, or mix the mains with your favorite flavor of FOH graphic EQ curve, monitor speakers live in the land of "flat is good." This is because any peak response frequencies in the speakers or mics will be the first to ring when the monitor is turned up. Additionally, monitors and mics have ever-changing responses when the two are positioned relative to each other. Then add "Tex" with his large brimmed hat, which makes a beautiful presence band reflector to guide even more wedge mix into his vocal mic.

FLACCID OR ERECT?

When it comes to mics, we all have been through the spiel about cardioid and hypercardiod mic pickup patterns, and where the feedback notches are. The challenge is to educate ourselves, and the performers, on how to position the stage monitors for each kind of mic. Cardiod mics like the ubiquitous Shure SM58 prefer the wedge front and center of the mic, with the mic normally bore-sighted from ball to cable at the wedge. I call this the "erect position." When dealing with hyper-cardiod mics like an EV ND767A, the wedge or wedges should not be bore-sighted but placed slightly offset. This is because of that nasty response lobe that super and hyper-cardiod mics have, especially in the troublesome presence band (2 to 8 kHz) that most mic manufacturers like to enhance. If you are required to center up a wedge on these kinds of vocal mics, teach your muso to position the mic horizontally (i.e., "flaccid"), to achieve the required maximum feedback rejection. From my experience working with performers with hormones, the terms flacid and erect get their attention, and the double entendre keeps it on their minds when handling the mic during the show.

EQUALIZATION

The whole process of ringing out is to counter the un-flatness of the contributing electrical signal processing and electro-acoustic transducers. That is why most monitor mixes employ electronic frequency equalizers between the monitor mix auxiliary output of the mixer and the monitor amp rack. Parametric or Graphic Equalizers are used for ringing out, with most monitor sound persons using a graphic equalizer. Lately, many of the new equalizer products are now introducing both kinds of equalization, especially graphic equalizers with high-Q notch filters that can be swept into place for surgical feedback removal.

One item to watch out for is the kind of graphic equalizer filters used for ringing. Both constant-Q and proportional-Q equalizers are useful, but you should be aware that different types of ringing require broader or sharper filtering. For example, the ever-popular Klark Teknik DN360 uses proportional-Q filters and makes wide, sloppy dips or boosts if used for only a couple dB. Likewise a Rane GE30 with constant Q filters maybe too brutal on the sound if severe notches are called for. Of course, feedback is not necessarily a high-Q phenomenon, and a slight low-Q notch may be the perfect fix.

BASIC METHOD

Ringing out the monitors works best with a few prerequisites. One is that your line checks are complete and all mics are up and close to their performance gain settings. The next is that performers have clued you in to their preferred mixes, and you have them roughly dialed-in. Then drop all your aux masters driving mixes to very low levels and kick on the monitor processing and amps. Taking one mix at a time, slowly crank up the aux master until a ring begins to appear.

Depending on your experience and style, you may take a notch-up or notch-down tactic in ring suppression. Experienced sound persons typically do a notch-down by hearing a ring, slightly drop the aux master level, drop the suspect frequency fader by 3 to 6 dB, and resume bringing the aux master level up to get to the next ring. This all works well if you have experienced ears and know your third-octave frequencies in your head.

Notch-up tactics take just a bit more time and work best with constant-Q graphic equalizers. The process is to hear the ring, back the aux master level down to stop the ring, and selectively push a suspect fader up until a ring occurs. If the ring frequencies match, then notch down that fader, and continue on. This allows the sound person to train on the frequencies at the same time as doing the ring out.

When about two to four ring tones are suppressed, that is about when you should stop the process on that mix and call it good enough. Other clues that you are getting to end on that mix ring-out are very little aux master increase as each ring occurs, or multiple rings at the same time. One other clue that something is terribly amiss with mic/monitor placement, or you suspect a bad mic, is when one ring frequency keeps coming up after killing it repeatedly. If you got anal about ringing-out, eventually your equalizer would trace out the combined frequency response curve of the wedge and vocal mic.

TRICKERY

One trick in learning to ring out is to employ a limiter circuit with an equalizer, either with an on-board limiter, or by chaining a comp/limiter in the aux drive lines, or by placing a comp/limiter in an aux send insert jack. Normally you would hardly limit aux send signals (if at all) to above 0 dBu, and likely in the +5 to +15 dBu range. By cranking the limiter down to about –20 dBu, you can start the ring out process and the limiter (or 10:1 ratio compressor) would catch and hold the ring at a low volume until you can search and destroy the ring with an equalizer. I call this the "training wheels" method as it allows you ring without the ring getting obtrusive or running away to extreme loudness, and giving you plenty of time for the frequency search.

Of course, other more obvious methods are out there to identify ring frequencies. Peavey's FLS (feedback locating system) is a patented display of LEDs next to equalizer faders that permits rapid ring identification and cues the operator for a fader notch maneuver. Similarly, Real-Time Analyzers (RTAs) can be attached to mics or signal lines to display rings or potential rings as they occur. In the sometimes boring part of monitor mixing, I can send a headphone PFL mix into my Neutrik Instruments Minilyzer RTA and track a mix in mid-performance. With very careful observation, I can pick out with my ears and eyes a trouble-making third-octave band long before it has a shot of blooming into a ring.

FINAL COMMENTARY

With the onset of more acts using personal monitors, the practice of ringing out will diminish, and becomes more a game of sweetening a mix with the performers. But there are still old-school, die-hard wedge users out there, and bands that have not invested into personal monitors.

One thing to be aware of is that many feedback suppressor products have entered the market with mixed results. While most get the job done, most do take some time to activate, so they'll let a ring persist for part of a second. The nasty thing about these automatic suppressors is that the detection algorithms are never good enough to determine a feedback ring from a loud guitar note held for a long time. So for you Carlos Santana wannabe performers, beware that the suppressors may engage if you push the guitar amp into the mix. Fortunately, the monitor mix sound tech is your best feedback suppressor, using all the computing power of a brain, eyes and ears.

One of the perplexing quandaries newbie live sound persons face is: "How do you know how to assign the channels?" There are many reasons offered, from many different sound persons, but most will begrudgingly admit to falling into lockstep with their mentors on this topic. Choosing instruments and assigning them channels can be done randomly, but more often than not there are reasons for following the crowd.

UP-DOWN, LEFT-RIGHT

When left to no other preferences, the traditional method to mixing console channel assignments follows two rules. The first is upstage to downstage (back to front) assignments. This usually means the rear located drum riser gets preference, plus any other upstage performers on risers. The next rule is stage right to stage left assignment, so that a FOH console has assignments that correlate left-to-right as the sound person looks upon the talent onstage.

TRADITIONAL THOUGHTS

Given this back to front, left to right sequence from the FOH position, several traditional groupings have come about from working with modern music groups. The first grouping, taking the first channels on the console, has been the percussion inputs. The second grouping is the bassist(s), typically composed of one or two inputs. The third grouping is the collection of keyboard inputs and guitar inputs. Typically assigned left to right visually, they may be grouped as keyboards and guitars in multi-act (festival) channel assignments to minimize the amount of preamp gain changes between acts.

The fourth grouping is usually vocals, and this tends to be placed in the middle to right half of the console, close to the subgroups or VCA groups. This location is meant to keep the sound person's work at the middle of the console, where riding vocal faders and subgroup/VCA faders are expected to be the main focus during the show. If there is to be open (unassigned) channels, they would tend to be in the keys/guitars grouping and the vocals grouping.

The effects returns are considered the sixth grouping, and they are placed either far right on the console, or on dedicated returns above the subgroup/VCA master section. Generally the effects are two or more stereo effect returns that are minimally altered during the show, at least in levels and equalization. The last grouping is comprised of the CD and tape playback inputs for intro/outro and break music between acts or sets.

MODIFICATIONS

Practicality sometimes interferes with neat arrangements of console channel assignments, and sub-snakes are a prime example of modifications in the traditional way of doing things. This all has to do with sub-snakes running to the backline and side of stages back to monitor beach, where the main snake patching is done. **Figure 31-1** shows a typical stage with sub-snake runs. While a backline subsnake can pickup drums and other upstage-center mic/DI inputs, big stages may have backline runs from up-stage left and upstage right that can confound things.

Of course the "patch-master" can always cross-patch the inputs back to traditional groupings, but there may be a good argument to straight-assign the sub-snakes across the console. For example, a 12-line sub-snake for the backline may contain the drums and chew up the first 12 channels on the main snake and monitor/FOH consoles. Then the stage right sub-snake gets the next eight channels, and the stage left sub-snake has the remaining four channels on the 24-channel cluster to the left of the consoles' master sections in a 40-channel system. The remaining channels (25 to 40) would then be left for a downstage sub-snake to pickup the vocals and the effects and CD/tape returns.

DRUM ASSIGNMENTS

Assigning percussion to the initial channels of a mixing console has always been the traditional process, with kick, snare, hi-hat, toms and overheads being that tradition. This order mostly comes from where you truncate channels on the assignment list as the venues get smaller or the available channels diminish. For example, small to mid-size clubs will not need micing of the crash or ride cymbals for most drumming styles. If available channels are in short supply, the toms will lose some or all mics. And if you ever got down to one channel, you would still attempt to reinforce the kick drum over any of the other percussion sources.

FESTIVAL ASSIGNMENTS

Most multi-act performances or festivals that share monitor and FOH consoles, will enact a "festival" system of assignments with 40 channel or larger consoles. Following the above mentioned order of sources, the first 8 to 12 channels will be assigned to percussion. Approximately the next dozen (i.e., 11 through 24) will handle the remaining instruments. Another eight channels (25 to 32) will normally handle the vocals, leaving the last eight for effects returns and CD/tape inputs.

While most festival setups will standardize all the inputs such as guitar1, guitar2, etc., across all the acts, sometimes smaller shows with a dominant touring headliner and local support acts will "spike" the consoles. Spiking is a term used here in the sense that the headliner's gear, mics, monitors mixes or anything related to the console is not to be tampered with af ter the headliner's soundcheck. Then the support acts must adapt by using the remaining channels for inputs, and live with the existing monitor mix equalization. Of course, this is "bad form" by the headliner's production crew, but the practice is born out of necessity of having the headlining act changeover happen quickly. With the emergence of digital consoles with global recall features, such spiking is going away.

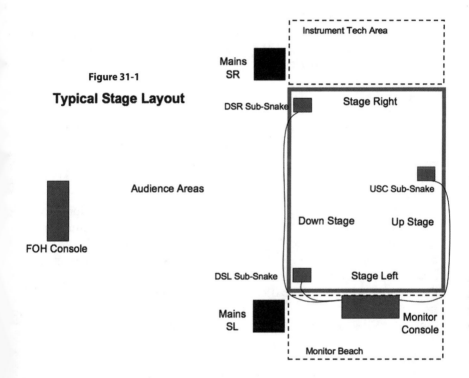

Figure 31-1

Typical Stage Layout

Mistakes while mixing is the human component of live sound production that we all are guilty of. However, with a little more attentiveness to details these minor blunders can be averted. This installment will recap common mixing blunders as a reminder to avoid them in the future.

We are not talking about gain structure issues, but having good gain structure habits is taken for granted. This means that 0 dBu is more than a good suggestion for average signal levels. Keep it in the green and out of the reds.

DOUBLE ASSIGNS

Of all the blunders I see, double assigning, or double routing, input channels to both master and subgroups is the most common. Only rarely should you need to intentionally double assign in order to boost signal levels when all other means are exhausted. But likely you have violated rule number one in mixing: "Turning something up can be done by turning everything else down."

So check before and during your show for channel double assigns. I typically run into this problem via two ways. One is the "fat fingers" resting on the faders and accidentally toggling the assign buttons. The other is from previous console operators, in which some have shied away from subgroups and assigned everything to the masters. It pays to always zero the console, even if you were the last person to use it. And if your assign switches also need the pan control to select subgroups, it is worth the half minute to run through all the input strips and verify all the channels are routed correctly.

MUTES

Rule number two in mixing is: "If it does not sound right, it is likely that something is wrong." This happens especially when accidentally having channels muted, and something just doesn't seem right. For example, I once was at a club watching the house tech mix the opening act at FOH before I mixed the headliner. The whole time he had most of the guitars and keyboards muted. But during the set he was happily adjusting the muted channel faders and tweaking the channel strip EQs, and using his "ears" to adjust to taste. All I could do is smile, and try to not deflate the house guy's ego.

When dealing with mutes and mute groups, get into the habit of not seeing red mute LEDs on active channels during the show. In essence, why not let a fader be all the way off instead of mute? Yes, there are pre-fader auxiliary issues to be aware of, but you can just mute out unused channels to keep you sane. And do not shy away from using a cue wedge or headphones to periodically check subgroups and the main mixes. Everything has got to be in those mixes to be in the mains speakers. After all, mixing with the stage wash is a lot less frustrating with all the channels unmuted.

COUNTER-EQUALIZING

This is another common blunder I see, especially with the headphone agnostic console operators. The classic case is the tech making a minor hack on the FOH graphic equalizer, and then proceeding to run through every input channel source on the console and remove the "hack" from the strips. Not only does this appear asinine from the outside observer perspective, but it takes away precious attention from the performance to run down the channels, knob-twisting away.

I watched an old soundman go through this counter-equalization ritual for the better part of the first set. Meanwhile he was missing every show cue to bump up solos and not keeping the lead vocal from being buried in the mix. It took every ounce of restraint to not reach in for the faders to hit the cues, while the soundman was busy trimming the EQ strips. I felt like a father teaching his teenager to drive, and hitting the phantom break pedal on the passenger side of the car.

The cure for the counter-equalization habit is to recognize the habit and to focus on the few input sources that need equalization, and not commit to a global equalization change unless it is well thought through. Channel strip equalization time can be done quickly just tossing on the headphones and PFL-ing the channel of interest. But do not spend significant time on this and lose track of the performance.

POLARITY SANITY

On low frequency sources like kick drum and bass guitar, take a quick moment to run through both options on channel strip polarity and stick with the loudest result. Many sound persons fail to at least do the checks, and just rely on the previous operator's opinion on polarity. Especially with different bass amp direct outs or DI boxes, bass instrument polarity can change from act to act. Do not accidentally fight the phase onstage, work with it.

SUBGROUP ASSIGNS

Much like the channel assigns, run through the subgroups and note both the assignment to the masters and master mixes. Some previous operators will hard pan pairs of subgroups, and your technique of mono-panned multi-subgroup categories will sound awkward if pushed to one side of the stacks. So do not forget those sub-group pan controls in the zeroing checklist. It is much easier to start mono on every subgroup before forming stereo positioned subgroups.

INSERTS ARE NOT INPUTS

This has been my mea culpa of late, as I have been accidentally routing effects returns and break music playback into TRS insert jacks instead of the TRS line inputs on the channel inputs. There is nothing worse than getting the "Where is the signal?" or "Where is that 'verb?" feeling when you fire up the system before the show. Or worse yet, having no effects in a time constrained changeover, and having to dive into the doghouse to check the patching when no signal shows up on the effects returns.

ZERO THE COMPS AND GATES

In the time crunch that always happens getting a guest console ready for your act, the insert compressor and gates are likely to get ignored until the show is under way. Suddenly the previous act's "squash the lead vocal like hell" settings become apparent in the mix. The blunder here is that you did not take the precious seconds to reset the comps and gates, or at least bypass them until you have the spare moment to set them up to match your signal gain structure. If you have to pull inserts on the fly, strongly consider dropping channel gain before pulling or bypassing the insert. Nothing says "dumb sh*t" worse than a sudden increase in vocal loudness when pulling an insert.

FINAL NOTES

Always zero up the console if you have the time to do so. If I am the first act of the night, I make it a habit to arrive early to get the consoles taped, channel assigned and zeroed. Even minor things like flattening up EQ strips before the show can play big dividends later on when a head-scratching feedback tone arises. Blunders are going to happen, but you can minimize them by thinking ahead of time. Even if you are a follow-up act, spending some precious time checking out the previous sound person's settings will give you the foresight in the triage of setting corrections you'll have to make at changeover.

Not meaning to brag, but the world of hum has been an absent partner from all my recent gigs. While I tackled hum back in my local soundco "good ol' days," I now realize how I was inflicting hum and noise problems onto myself. This diatribe should give you the tools to bust the demons of hum in your system, and let you know how to take preventive measures to minimize its occurrence.

PREVENTIVE MEASURES

Most hum problems revolve around cheap cable and gear, cabling and gear in need of repair or power distribution faux pas. The first rule in knocking out hum is the proper usage of balanced cables. While not a 100 % cure-all, twisted pair wiring with a shield is a great start. This means XLR to TRS interconnects throughout. With the cost of balanced patch cables not all that much more expensive than unbalanced cables (signal/hot and shield/cold), using unbalanced cables at any primary signal path is just foolish.

Figure 33-1 shows a typical electrical schematic diagram with a microphone source to mixing console to drive signal processor path. Smart persons will indulge in balanced cabling between each piece of gear. The cold wire in the console to processor interconnect is in red in the diagram, so you see what the difference in balanced to unbalanced cable paths is. In balanced connections, the hot (XLR pin 2) and cold (XLR pin 3) carry the signal in out-of-polarity forms. If hum currents couple via the shields or grounds, both the hot and cold wires get nearly identical hum noise on the wires. Each piece of gear receives balanced cable signals and does "difference amplification." This means the hot

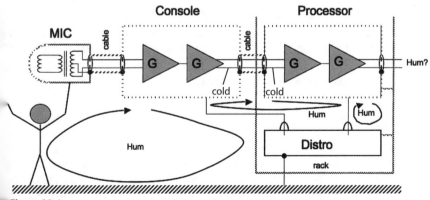

Figure 33-1

signal and noise is subtracted from the cold signal and noise, and differential signal from the two wires remains.

Because the hum and noise are common to both hot and cold wires, the differential amplifier subtracts the signals, resulting in the cancellation of the hum and noise, leaving twice the individual wire signal as the output. So not only are you busting hum, you're getting a higher signal capability too. And if you are ultra-cautious like I am, almost all your balanced cabling is configured as "star-quad" type.

Star-quad balanced cabling uses redundant hot and cold wires for a 4-wire plus shield cable design. The hots and colds are interleaved to further reduce coupled hum, audio noise and radio-frequency noise on the transmitting and receiving pieces of gear. Although terminating star-quad cable with XLR connectors is more of a chore, I do it for additional peace of mind.

Now back to **Figure 33-1** again. If that cold wire were missing, as in the case of unbalanced patch cabling, the shield serves as the cold path and ground reference. Because the shield also has hum and noise currents from power supply redundant paths (rack chassis, adjacent processor chassis, distro branch circuits, etc.), all the hum is injected into the signal path and not rejected at the difference amplifier.

GETTING CHEAP

And why did you use unbalanced cables? Because you were cheap, did not understand the benefits and the cables were handy. Short 2-, 6- or 10-foot XLR patch cables are not something most of us have in excess just hanging around the shop, and we generally have to make the time to solder-up these specials. So most hum problems are created by just not knowing the advantages of balanced signal transmission, or not having the handy balanced patch cables to make it happen.

Getting cheap also extends to the gear you are using. While most live sound gear can receive and transmit balanced signals, some pieces of make a half hearted attempt at it. Servo balanced transmits and differential receives are the expectation, but sometimes gear is provided with "impedance balanced" transmit (output) circuits. In this case, the cold wire connection is just terminated with a resistor to ground, instead of a proper drive circuit. While it is better than a grounded cold wire, the ability to reject hum and noise is less than servo-balanced drive circuits. So when choosing your next purchases, note the balanced drive circuit types.

AFFLICTED

Even with the best intentions, you may be doomed to battle the appearance of hum in your sound system. Your best bet is to dust off your troubleshooting skills and start the divide and conquer process. Typically my routine is to check the console inputs first to see if I have a single input hummer. For example, a recent hum intrusion for me revolved around the bass guitar direct input. After muting the channel strip and assuring that I had isolated the culprit, the next step was to go on stage to troubleshoot further. Upon checking for worn cables and DI issues, I found the problem in that the bassist had plugged his amp into a forbidden receptacle, instead of the handy stringer receptacle box I had conveniently located on the floor near his amp.

If your hum is not at the console, but further down the drive line, two items should enter you mind. First, worn or broken wires in cables; and second, differing sources of power to the power amp racks and the drive racks. Simple troubleshooting is again the obvious choice. Turning the humming power amp's attenuator down should drop the hum. If not, the amplifier or amplifiers are the issue. Going back towards the console, cabling can be checked by dropping amplifier gain (increasing the attenuators) and opening the patch cables and snakes along the way. With balanced cabling, an open balanced connection should result in very quiet power amplifiers upstream. On the other hand, a half busted balanced wire or an unbalanced patch can be a very noisy affair.

If you have hum problems, and you have to live with distro connections or other bad signal flows, you should have a last ditch solution in the form of isolation transformers. Isolation transformers break the pin 1 (ground) connection and insert an audio transformer into the hot and cold connections. The industry standard for isolation transformers is the Sescom IL-19 that comes in a handy barrel-XLR format to break into XLR patch cables. Because of its small size, the smaller internal transformer can only handle modest line signals without distortion. But I will take lower drive levels over hum any day. And serious sound companies always have a pair or more of IL-19s (or equivalent) stashed away for Just-In-Case usage.

I hope this provided some common sense notes on hum busting when under time pressure of the gig, or when configuring systems back at the shop. If you are taking out a new batch of gear for the first time, consider doing a "dry-run" setup back at the shop first. Yes, it does consume time — but I would rather tackle hum and noise problems at the shop than last minute at the gig. Remember, it is the sound company's duty to put the Boy Scouts to shame with their preparedness.

I believe the role of road cases is critically important to the profitability of a sound company. But not every case made out there is really road-worthy. A lot of my anklebiter friends think I am insane to pay $600 or more on semi-custom rack cases to haul power amplifiers and signal processing gear. I look at it as minor insurance to keep the gear performing throughout its scheduled life — not to mention the benefit of putting the gear on wheels instead of having my back do the lifting. This installment is about rack road cases, and a guide to getting them built and maintained.

RUGGEDNESS

I am sorry to say that if you are going to the music store or to a typical pro-audio store to buy a rack case, you are falling prey to the idea of road cases, but not the execution. With the exception of some smaller ¼-inch wood signal processing rack cases for guest engineers, all the lesser shelled cases are just junk chasing your foolish dollars. I believe we all go through a phase with ¼ -inch plywood or blow-mold plastic rack casing while learning the ropes, but sooner or later you are going to have to bite the bullet and move up to professional ³/₈- or ½-inch plywood under laminate casing. Yeah it is heavy, but behind all the wood and aluminum extrusions is a rack case that'll be around for more than a decade of road abuse.

SHOCK-MOUNT OR PULL-OVER?

All professional rack cases are shock-resistant to certain degree, but rack case designs typically come in two versions: shock-mount or pull-over. Shock-mount rack cases look like music industry (MI) rack cases, but have a surround of foamed rubber a couple inches around, with an inner rack and an outer shell in between. Typically the bottom foam rubber will be stiffer in order to handle the weight of the inner rack and gear attached. And shock-mount rack cases will be obviously a little larger in dimensions, and have less rack space efficiency inside.

Pull-over rack cases do not have front and rear lids, but have a single latched seam at the bottom to come apart as a wheeled tray and a pull over top that hides the inner rack during transportation. Like shock-mount rack cases, the tray has stiffer foam that the inner rack rests on, but the top's interior has a softer foam inner lining that fits nicely with the inner rack. The inner rack may also be laminate based wood and aluminum construction or epoxy coated wood.

RACK RAILS

The heart of the case is front and rear rack rails. These L-shaped steel pieces with 10 – 32 taped holes follow the nearly century-old telephone company specification for mounting 19-inch wide rack gear in 1.75-inch height increments. Most professional racks will have a modest 1- or 2-inch rail recess from the inner rack wood or aluminum for clearance of control knobs, switches and gear handles.

Also, there may be one or two sets of rear rack rails for rear attachment, as heavy gear needs some of its weight supported to keep from bending up the faceplates on the front rack rails during rough transportation. Typical front to back rail-to-rail spacings are 11, 14 and 18 inches, depending whether you have signal processing or short or deep power amplifiers to contain within the rack. A second, more rearward, set of rails maybe installed for connector panels or fans.

TRUCK PACKING

Most professional rack cases will be accessorized with hardware to facilitate the packing of many rack cases side by side. Little things like recessed handles and latch plates (spring-loaded) take up less air once inside the truck or trailer. And having a ¾-inch or heavier wood caster board on the bottom of the rack case is a necessity for case and wheel attachment. Well-designed caster boards should also have pairs of hand-grip holes on the edges for roadies to lift the rack case when necessary.

Casters should be the 4-inch (or larger), rubber-wheeled types. I believe most of us know the thrill of dragging heavy rack cases through soft sand or small gravel, and it is next to impossible with smaller wheels. You do not have to buy the nice Guitel brand blue swivel caster wheels, but the $15 to $20 apiece you spend on them is really worth it. If your truck or trailer is not equipped with sidewall strap features to hold the cases captive during transportation, then I recommend the two rear casters be upgraded to locking types for movement minimization on the road. And locking casters are also handy when they are placed on un-level surfaces or used as step-stools.

COLOR MY WORLD

I hate to say it, but I agree with Henry Ford when it comes to color. Satin Black (between gloss and flat) is my choice in laminate color when it comes to rack cases. Not only does it hide dirt and dust, but it gets back to stage psychology. Anything black tends to be ignored by human eyes. Now I understand you may choose a different primary color for finding your cases during festival shows, but in corporate gigs, black hides better backstage from a show planner perspective.

To maintain rack cases, a little periodic dusting with a shop rag goes a long way. Spiders and other critters just love collecting below the caster boards if you give them a few days. And I recommend a yearly treatment with the roadie's secret weapon, WD-40, to maintain the satin look and beat back corrosion on the steel and aluminum hardware. Do not lube the casters unless they squeak, as every bit of oil or grease lubrication will collect dust and dirt. Have a can of silicone or graphite spray for lubrication if possible.

BUILD OR BUY

I recommend everyone should at least attempt to build some kind of road case, just to get an idea of the time and effort required. You'll either enjoy it and take road case building as a sideline, or get wise and pay someone handsomely for their expertise. After building a pair of drum hardware cases, I quickly learned that the amount of parts scrounging, gluing, screwing and riveting was not for me any more. But neither were the cheap sh*t cases coming from the stores.

To find a good road case builder, do a little networking and find what other soundco's are using. There are the national brands, some which advertise in *FRONT of HOUSE* magazine, and smaller road case shops that market by word of mouth. Save the discount cases for the musicians, and buy quality rack cases with ½-inch or better wood construction from good case builders and you may only have to buy casing once (buy once, cry once; buy cheap, cry twice).

ADVANCING THE GIG

October 2005

When you contract for a new gig at a venue you never have been in before, checking out or "advancing" the venue weeks before the gig is a very good idea. By knowing what you are logistically facing ahead of time, your truck pack is optimized and you won't be bringing the kitchen sink to cover the unknowns. I'll run down the aspects of advancing the gig here to set expectations. Then I'll take the middle of the road example by checking out a typical large club venue.

COME PREPARED

After you have directions to get to the venue — and, most importantly, know that the venue is open for inspection — there are certain little items to carry or stuff into your pocket before walking in. The most important thing is to have a paper and pen to record your notes. Your memory cannot retain the details after advancing dozens of venues; after a while the rooms all blur in details. I keep a Daytimer zip folder, as it contains paper, pen and all my essential contact information.

Next, having a couple of your business cards and the act's stage plot connotes a touch of professionalism when introducing yourself to the room's management. Other handy items are a small flashlight and an electrician's cricket. Many times the stage area is poorly lit or completely dark during the off-hours,and a flashlight helps get the essentials identified, and having the cricket is a nice non-contact tester to see if there are hot power receptacles and bull-switches located near the stage.

CHAT 'EM UP

When entering the venue to be advanced, find an employee that looks to be in charge, and warmly greet them. Introduce yourself, tell them who you are, what production company or act you represent, and what your objective is being there. Then go about your wandering notating business. Many times you will find that the venue staffers will summon the manager, or provide on the spot answers to your questions as you wander through the stage, dance floor and FOH position. Do not just walk in and anonymously start taking notes, as you could be mistaken for a health inspector, tax assessor or any other official that may get a less than friendly greeting.

If you can, arrive in the afternoon on a weekday so that no other production or band gear is likely to be onstage. This timing also is more conducive for managers or staffers to answer your questions without being pressed for time. After your venue inspection, be friendly and ask good questions, both technical and non-technical. Knowing how the venue's usual crowd breaks down in such areas as size, age and preferred music genres is just as important as the room's size and acoustics. And knowing things like set start/stop times and when the place fills up are important in executing the show.

THINGS OF NOTE

The following are the items to note when advancing the gig. First priority is the size of the stage, stage height, work light switches and covering materials of the stage area. Sketching a floorplan, with dimensions and material notations, is essential. Since most of us know our own height or have good idea of about how big 4-foot by 8-foot sheets of plywood are, you can make pretty good judgments in stage size, dance area, size and distance to the FOH position.

Next, sketch the load-in door location, and count any stairs to be overcome at load-in. Also, notate the preferred production vehicle parking and unloading locations which you should have gotten during part of the management dialog. And, when noting the FOH position, record any obstacles to getting consoles and racks into the position.

Besides noting materials for room acoustics (clap your hands to check decay times and warmth), give the venue a quick chair count. Knowing the seated capacity of the room will give you a logistical cue as to how popular the place is and a feel for the kind of money the venue will retrieve if you pack the house. It is something to consider, especially if you get paid $500 but the venue has 200 seats with a $5 dollar cover charge.

When checking out the stage area, inspect the quality of the electrical service. The difference between some wall receptacles and a 100 Amp, 240 VAC disconnect switch is crucial. If in doubt, ask the manager or staff how many circuits cover the stage receptacles. Noting light bar hangs and hooks for overhead routing of cabling to FOH is important.

After catching all room aspects, come back to the venue personnel to make your final chat. Do not forget little items like the typical age of the crowd, when the venue usually packs in and typical genres of music expected in this venue. These important demographics will help the act tweak their set list. Ask about normal load-in times and preferred entrance doors for musicans and production. Some venues will not allow load-in at dinner hours — or worse yet, until after the dinner crowd rush. I have personally been requested to do an 8 p.m. load-in for a 9 p.m. show, and it's no fun.

BAND ENGINEERS

If you are a band engineer going into a club install, you have the additional task of inspecting the production gear, and noting what is and what is not provided for your show. This may prompt you to prepare your own outboard effects or other essential signal processing needed to pull off the act's show. Make sure you leave with a stage plot, and get the house tech's phone number to chat him/her up as well as to set expectations.

RIG SQUATTERS

Another practice that is increasingly becoming prevalent is the act having to pay for the install production crew and for gear rental. These rig "squatters" arrange with the venue management to leave their gear in the venue, but will remove it or relocate it if an act insists on bringing in its own production. Generally a middle ground arrangement exists, where the band engineer operates the install gear, but the act pays for the install gear rental instead of hauling in their own for production. These rig squatters exist because many acts have poor or non-existent production support, and are willing to pay for the convenience.

It is important to find these rig squatter situations in advance, so that on the eve of the gig the act is not unexpectedly hit up for the gear rental and technician fees. For example, an act may get an $800 contract for a show, but in the fine print the rig squatter will demand $200 for the gear rental and another $100 for the rig technician off the $800 gross pay if you do not inform the rig's owner that you are providing production. If the rig is worthy, it may be worth the $200 to not hump gear for that gig. But be sure to notify the rig owner to leave the tech at home to save that fee. Also, with these squatters you may have to "qualify" with the rig owner before they will let you run the gear without a hired tech.

In Chapter 26 I addressed this topic, but took a historical and Q&A approach to justify things like 0 dBu levels, gain/level settings and +22 dBu brick walls. When I first published that column in *FOH*, a lot of readers needed to get more into the real knobs and faders portion of setting up their consoles. This chapter will revisit gain structure and do it in a straightforward way.

ALL THOSE CONTROLS

As an analog mixing console fanatic, the sea of knobs frightens a lot of newbies, and begets the oft-asked question, "Do you know what all those knobs are for?" The answer should be "Yes," and most key knobs are the gain controls for the mic or line pre-amplifiers. Most have a 40 – 70 dB range of gain adjustment, and are meant to bring -60 to -30 dBu microphone levels to line levels. I mention these levels as a range, because singers vary their acoustic input to the microphones greatly. So what starts out as gain adjustment for a subdued lead vocalist at soundcheck, will likely be pegging past +22 dBu after the preamp during the last set. But by trimming the soundcheck mic gain levels at 0 dBu peaks means that I will have 22 dB more headroom for that vocalist to blow his/her lungs out later at the gig.

But if I get rational vocalist or musician who soundchecks at performance volumes, I will dial-in their mic gains for 0 dBu averages and 6 to 9 dBu peaks. The reason for this is that I expect the levels to stay about the same through the performance, and I plan to have the channel fader at 0 dB when that channel is the prominent feature in the mix. This allows me another +10 dB of gain increase if I need it at the fader, and will likely not clip out the channel circuitry. And of course, the plan is to operate fader usually lower than 0 dB when the instrument/voice is supposed to be in the background in the mix or, worse yet, when the wash coming off the stage to your mix position buries the fader to nearly the off setting.

THE BASS-ACKWARDS METHOD

Occasionally, you will see an anal-retentive mixing console operator who starts with all his/hers channel faders at 0 dB, and then begins adjusting the gain controls from the lowest settings to the final setting. Now, this is okay if the final settings are about 0 to +10 dBu on the peaks, but you'd still expect the operator to begin to use the faders to mix the band at soundcheck and show. But normally these bass-ackwards operators will mix using the gain controls on each channel, leaving the faders in a neat and pretty 0dB row. These are also the sound persons you want to avoid, as they have egos as large as army tanks, and think their way is the only right way to mix.

WHAT THE DESIGNERS INTENDED

As an electrical engineer and former console designer myself, the mixing console design priority goes to the microphone preamps first, and not at the fader control electronics. The reason is that the best low noise performance is made at the preamps, and the successive circuitry can be noisier without much impact on the whole console's performance. To grade a preamp, the term Noise Figure is used to describe how much loss of Signal-to-Noise ratio (SNR) a preamp contributes to the pristine inputs signals. To characterize both the mic preamp and the rest of the mixer/summer stages as Noise Figure, the following equation is used:

$$F_{console} = F_{preamp} + [(F_{summer} - 1)/G_{preamp}]$$

From the above equation, the overall Noise Figure of the console is the preamp Noise Figure plus the mixer/summer Noise Figure divided by the preamp Gain. What you should conclude from this equation is that strong preamp gain settings do the best to reduce overall system noise, and that using the faders to mix has less impact on noise than mixing from the gain knobs.

CROSSING TO THE OTHER SIDE

As in the previous gain structure chapter, there are a lot more methods to sorting out gain and level settings once the signals leave the channel strips. Console designers start introducing attenuation strategically at the summing amplifiers for the subgroups, auxiliary masters and Left-Right-Center masters so that multiple high level signals do not statistically sum to a clipping level. So a summing amplifier may lose a couple dB before the fader control. And the fader control loses 10 dB while actually indicating unity gain (0 dB), so that the follow-on +10 dB make-up amplifier can provide the optional gain boost before hitting the output connectors, metering or next summing amplifier stage.

From my experience, it is best to kick off the show with the vocal subgroup at the nominal level (0 dB), and place the other groupings a few dBs behind it. Then, at show start, you can focus on vocal loudness with the master LCR faders, nail the mix loudness within a second or two, and then begin to back or forward the non-vocal subgroup faders to rough-in the mix. Finishing touches can be done using the channel strip faders, starting with key instruments like guitars, then bass, then keys, then drums. Most times I will have the channel faders rough set from experience with the act onstage, so I am not at 0 dB on every fader, scrambling to immediately lower half the channels at show start.

For most rock/country/pop acts I do, there is one lead vocalist and the rest of the vocalists are backing. For this configuration, the lead vocalist is pegged at 0 dB with all other vocals starting at -5 dB. This way I can still place the lead vocal on top of, or just within, the other instruments, and use the subgroup and master faders to set the overall level of the show. If other vocalists switch-off taking lead vocal duties, everyone goes to -5 dB until the lead vocalist is known (hopefully before they start singing, but that is another matter).

DRIVE CHAIN

Opinions vary at the at the master LCR faders. I am a devoted -5 dB to -15 dB master fader guy, as I like my master metering to barely bubble beyond 0 dBu depending on the venue and stage wash conditions. But if you like backing off your power amplifiers by about 10 dB, then you are running close to 0 dB on the master faders and living on the threat of having only another 10 dB of console "gas" before clipping console circuitry. The lower master fader levels are a hallmark of us old farts who lived and died by VU meters and amplifiers with only a couple hundred watts of amplifier power to perform on. Today amplifiers require +4 to +10 dBu of input signal, but they can knock hundreds to thousands of watts into the mains speakers at those levels.

If you go the old fart way, you may have a touch more hiss in your speakers than if you trim back amplifier level/gain controls. But if you do run attenuated at the amps, keep a watchful eye on your console, equalizer and speaker processor levels for clipping or limiting.

If you follow my way and leave your amps wide open, make sure your speaker processor hits limiting at or just before the amplifier delivers full un-clipped power. The limiter settings are generally chosen by either speaker driver maximum program power rating, or amplifier power ratings, whichever gets in trouble first. For example, a 1200 watt amplifier that clips at +7 dBu may deserve a processor limiter setting at +6 or +6.5 dBu. And some of the better processors provide both compressor and limiter settings to better tailor the driver power dissipation characteristics as a measure to prevent voice coil destruction.

CONCLUSIONS

Please use this discussion as "food for thought" piece, as there is no absolute way to get gain structure perfect. I encourage you to comment on your methods via e-mail or on the Wild and Woolly forums at FOHonline.com, to share the expertise.

Every soundco has cables, endless numbers of them — from snakes to mic patches, to speaker cords, to AC power distribution. And like most newbies, soundco's start with the usual "whatever is handy" approach to transporting them — old road cases, milk crates, suitcases, steamer trunks, plastic totes, etc. This article describes the art of cable keeping and how the big soundco's do it.

Sometimes we never learn, but typically plenty of clues are dropped by persons witnessing your load-in ritual. Comments like "How many trips are you gonna make?" or "Hey, it's getting cold in here, wanna hurry up?" should tell you are not very efficient in getting your gear from the rig into the venue.

The natural way to resolve this is to consolidate those suitcases, tubs, whatever, into bigger storage items. And the natural path to this resolution are work trunks, work boxes or work cases. Whatever you call them, it means cases on wheels that carry all those cables for fewer trips in and out. And you do not have to go my way with custom cable trunks, as many case makers and pro sound retailers offer standard size trunks that are nothing more than spaces to fill with the necessities of gigging.

For example, the two figures in this chapter are cable trunks built for me specifically for stowing cables and related items. For my "C" rig, I have three such cases with nominal dimensions of 22 inches by 44 inches by 22 inches in height. They all have dolly boards and caster cups for stacking, although with cables it would take two to four roadies to stack them. All these cases are constructed with lower wells and two upper trays for smaller items.

Cable Case One has slotted sides to the well for main snake storage, and the trays mostly hold XLR patch cables and sub-snakes. Case Two is everything AC, with bigger 12 gauge AC cables in the well and short patches and quad box breakouts in the trays. Case Three has a well divider in the long dimension for keeping K&M tripod mic stands on one side, and Speakon patch cables on the order side (kept in order of length). The trays on case three hold mic pouches, extra mic clips, windscreens, DI boxes and the beloved muso convenience store inventory (spare drum sticks, drum keys, guitar and bass strings, batteries, guitar cables). And the whole mess moves in using only three trips for one person.

CABLE ROLLING

For smaller cables like XLR patches, I teach my cable rolling helpers to coil in 7-inch circles using the over/under technique. The technique is better shown than described, but uses a lot of common sense and a bit of Navy seamanship. Generally you attempt to get the cable loose from all others and in a somewhat straight line. Grab one end with your holding hand as the other hand pulls about 2 feet of cable back to the holding hand. As you coil the cable, you alternately twist the cable one way and then the other, with your thumb and fingers making cable half turns as each coil is formed. You will notice that the far end of the cable will not be twisting if this is done correctly, as it prevents loops and tangles.

Now on to the subject of cable ties. My cable ties have been the evolution of my rolling cables since the early 1970s. Back then Velcro was still too new, and tape left way too much residue. And if you knotted your cables, you quickly found out they would not last very long. But being of practical means, and also an office janitor between gigs, I had a ready supply of yellow Glad garbage bag plastic ties — perfect for mic and guitar patch cables. As my occupations improved and my source of free cable ties dwindled, I switched to Velcro ties. Still, the yellow Glad ties were easy to spot after gigs, and had reasonable life spans — even today I still have a few in use.

Today I have given up the fancy cable ties for "gecko tape." My half-inch yellow gecko tape is still made by Rip-Tie Inc., but comes in 150-foot rolls with hooks on one side and loops on the other. By cutting 6-inch lengths of this tape I have cable ties that cost barely a nickel each and last through dozens of gigs. I make it a habit

Figure 37-1

to stash all ties near the snake patch box for easy tracking at the end of the night. **Figure 37-1** shows my 7-inch coils of XLR patches with yellow ties in the trays.

TAMING SNAKES

Coiling up snakes is straight seamanship, just figure-eight the snake in the cable trunk, leaving about 10 to 20 feet of patch box hanging out until the rest of the snake is stowed. **Figure 37-2** shows my less than perfect figure- eight stowage of the snake cable. By using the figure-eight technique, no tangles or twists occur when extracting or retrieving the snake cable. The extra bit of patch box and cable lets you store the case at monitor beach, and pull just the required amount of snake to FOH, leaving the rest stowed neatly.

AC AND SPEAKER CABLES

These beefy 12 gauge cables with black jacketing get a similar treat-

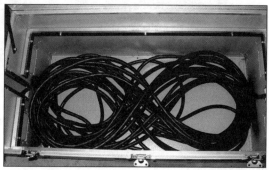

Figure 37- 2

ment as my XLR ones do, but in a 12- to 15-inch coil size. But I have a little tip to share with you: If you can get some theatrical black trick line, cut healthy lengths of it (about 2 feet) and square knot the middle of the trick line on one end of these kinds of cables next to the connector. If you can tie your shoes, then you can tie off the coiled cables quickly and toss them into your cable trunk. If trick line is not handy, then black boot laces make a nice substitute.

38 INSTRUMENT FREQUENCY

February 2003

When putting together a mix, it is handy to have the knowledge of the kinds of sound sources and the expectant frequencies generated by each. This chapter is intended to be a quick reference to frequency bandwidths of various modern music sources and a table.

DRUMS

Starting with the traditional left side of the console, drums are usually tuned to resonance from about 160 Hz to 800 Hz. But the total bandwidth of each drum can range from two octaves below that, from batter head sounds (click), into the presence bands (2 to 8 kHz). For example, a 22-inch kick drum batter head is typically tuned to E3, or 164 Hz. But sub-harmonics are given off at 82 Hz and 41 Hz, these are the chest "thump" that should be felt more than heard. And with the second harmonic suppressed (328 Hz) to make room for other instruments, all that is left is the "click" around 3 kHz.

Other drums are typically tuned a bit higher: a 16-inch floor tom resonates at C4 (261 Hz), a 14-inch floor tom at F4 (349 Hz), a 12-inch rack tom at A4 (440 Hz), a 10-inch rack tom at D5 (587 Hz) and a snare at G5 (783 Hz). If you set drum gates, then the first sub-octave below the batter head resonance is usually where the frequency band is set. From the above example drum tunings, the kick is at 82 Hz, the 16-inch floor tom is at 130 Hz, the 14-inch floor tom at 175 Hz, the 12-inch rack tom at 220 Hz, the 10-inch rack tom at 293 Hz and the snare at 366 Hz.

CYMBALS

The "sizzle" in most cymbals ranges from 2 kHz to 40 kHz, but us Neanderthal male soundmen will struggle to hear beyond 16 kHz. If you want to catch some of the "impact" of crashing cymbals, then adding an octave or two below 2 kHz (1 kHz, 500 Hz) is a good place to start. The tough choices come with mic placement and bandwidth allocation, to choose the amount of separation between drums and cymbals. If you are just doing kick drum and overhead condenser mics, then 100 Hz to 16 kHz wide open overhead mics are the ticket.

BASS GUITAR

Bass guitars of the four-string type range from 41 Hz (E1) to about 246 Hz from a fundamental note range. Adding the necessary octave or two harmonics should provide at least up to 1 kHz on the top end, and more if some finger sounds are desired. For 5-string basses, the low B-string at 31 Hz (B0) means that some really enormous subwoofers are required to faithfully reproduce the instrument. Fortunately, we humans can catch the sensation of low frequencies by hearing the second octave harmonic (62 Hz) and the higher harmonics — but this means that your sound system needs to be still flat at 62 Hz, and not trailing off by several decibels.

GUITARS

Guitars follow the bass guitar, but at an octave higher. So 82 Hz (E2) to 659 Hz (E5) are the fundamental frequencies. But with electric guitars with fuzz boxes or overdriven amplifiers useful harmonic outputs can extend up to 5 kHz and beyond.

KEYBOARDS

A full 88-key piano ranges from A0 (27.5 Hz) to C8 (4186 Hz) plus harmonics for a very broad bandwidth. And the classic Hammond B3 organ runs from 32 Hz rumbles to 5920 Hz whistles in blues and rock usage. Then with synthesizers, all semblance of audio noises are possible over the audio range and beyond. With human hearing in the 16 Hz to 20 kHz range, that is a reasonable bandwidth limit.

OTHER INSTRUMENTS

Some other typical modern musical instrument frequency ranges are the tenor sax at 110 Hz to 587 Hz fundamental notes. Violin or fiddle fundamental notes range from 196 Hz to 2093 Hz. And blues harps can range from 196 Hz to overblows at 2959 Hz. Note that you should add a couple of octaves beyond the fundamentals to catch most of the nuances of these instruments.

VOICE

Just like instrument is, human vocals are given the Italian music prefixes of baritone, tenor, alto and soprano designations. Fundamental note vocal ranges are baritone at 110 Hz to 392 Hz, tenor at 146 Hz to 440 Hz, alto at 196 Hz to 698 Hz and soprano at 261 Hz to 1046 Hz. And since the voice has so many sounds beyond the fundamentals, having good vocal reproduction to 8 kHz and beyond is a good idea. Shaving off above 12 kHz is a good idea when the vocal has plenty of cymbal leakage in the microphone.

While this chapter is not theory, it is a whole lot of practice, because a contractual technical description of your provided rig is an important part of winning bids. Yes, you can call contract terms part of the "Biz," since most of the smaller gigs you take do not include riders and just want you to deliver the sound (and sometimes stage lights) in a professional manner. But by describing gear in quantitative and generic ways, you concretely express to your customer that you are all business and not over-promising a touring rig for an amateur event.

So first let's show by example, with a typical small soundco technical contract section for a typical small indoor or outdoor event. Below is Exhibit A, the provided equipment list:

EXHIBIT A
Equipment List:

The Contractor agrees to furnish a COMPLETE PROFESSIONAL QUALITY sound reinforcement system for audience and stage. The Contractor will provide a venue-appropriate system for live music reinforcement. The Contractor shall provide a minimum of:

- One _____ XLR input mixing console of professional quality
- Speakers, Amplifiers and Crossovers as appropriate to venue and music style
- _____ subwoofer cabinets
- _____ top box cabinets
- Signal processing and amplification appropriate to above cabinets
- Equalizers of professional quality (1/3 octave capable) for _____ monitor feeds
- _____ insertable Compressor/Limiters for vocals and instruments as required
- _____ Digital Reverb/Effects units for vocals and instruments
- _____ monitor mix feeds to the stage
- _____ monitor wedges (12-inch+horn) for the performers
- Up to _____ mic stands and booms, and _____drum mic claws
- Mics –Assorted professional-quality mics of good working condition
- _____ Direct Insert (DI) box (passive) feeds of good working condition
- Cabling as required for operation of above sound equipment

Lighting requirements. The Contractor agrees to furnish a COMPLETE PRO-FESSIONAL QUALITY lighting system for the stage.

- _____ PAR56 300 Watt MFL lighting fixtures with basic gel colors in frames
- _____ stage lighting lifts with bars supporting the fixtures
- _____ digital dimmer channels for the fixtures
- A digital control console for remote control of dimmers
- Cabling as required for operation of above light equipment

GENERIC DESCRIPTION

Note that in the above **Exhibit A** narrative the brands of gear are not listed, but the kinds of gear are listed in a way that indicates you are bringing a sizable amount to do a professional show. This leaves plenty of leeway so that you do not get caught in haggling over a certain price or whether a brand of gear is professional enough.

For example, a regional sound company could use the above equipment list and bring the Midas console, VerTec mains and subs and a pile of Clair 12AM wedges. At the same time, the equipment list could be a vintage Soundcraft Delta console, JBL SR tops and subs and Yamaha club series wedges. Obviously a big difference in size and capability, but the bottom line is that you promised a "venue appropriate" system, and you made the call on appropriateness.

MORE POWER, SCOTTY!

In my generic contract, I also have an **Exhibit B** for power requirements. To me, solid electrical power has to be a given in order to pull off a successful show. Below is my **Exhibit B**:

EXHIBIT B
Electrical Requirements:

The Buyer shall provide at minimum, a 100 Ampere 120/240 VAC single phase or 100 Ampere 120 VAC three phase electrical service for show performance. The service point shall be within 100 feet of the performance area, be available during the time of load-in through load-out and be up to current National Electrical Code (NEC) standards.

The service point shall be at least four-wire (hot, hot, neutral, ground), and disconnectable by means of either a plug-and-receptacle or through a company (bull) switch. The safety ground shall be a single point bond at facility electrical service entrance, and shall not carry neutral current at the point of disconnect.

The preferred methods of service are a 100 A or larger sub-panel, or via 100 Ampere or larger company switch capable of accepting #2 gauge or larger feeder conductors.

Alternatively, portable generator service shall be not less than 48 kVA rated and grounded by the generator rental provider. The generator shall be operational and fueled to be able to operate for the whole duration, from load-in to load-out. The generator shall provide stable 120 VAC and 60 Hz currents to prevent equipment damage. Generator location should within 100 feet of the performance area and secure from unauthorized personnel tampering with generator or tripping on the Contractor's feeder cabling from the generator to the performance area. Whisper-quiet (show power) generators are recommended. No vehicle traffic permitted on feeder cabling.

The Buyer shall be responsible for all Damage to the Contractor's equipment should the electrical service fail on-site.

Now I do not expect perfect power and shiny new connections, but the electric utility does promise the venue clean power, and I just want what they are giving. Bad power from in-venue noise sources and poor wiring can make the best touring rig sound like crap.

Note that I also talk about portable genny power. I cover the ups and downs of good generator selection in other chapters, but putting the desires into words in a contract exhibit helps a lot. No sense getting surprised by somebody's RV generator being left near the stage as the sole source of juice for the event.

OTHER CONTRACT ITEMS

While the technical items in the contract are the focus here, do not forget the basics of contract provisions — things like the itinerary, payment terms, authorized production personnel, security, insurance, permits, parking, unions and inclement weather clauses. There's a MS Word version of the complete document placed on the fohonline.com/contract Web site for all to ponder its contents, and you can edit a version of your own.

Obviously, not all production buyers will go along with all the provisions of this contract — it should serve as a starting point so that buyers and contractors can mutually agree to line-out and initial the changes. Note that I did not include the typical artist vanity requests like catering, fresh linens (towels) and break room. I will leave it to you, the contract writer, to have the cajones to include those provisions. Iced bottles of Evian water, M&M candies with the green ones missing and shower facilities are things of soundco crew fantasy.

Back in the early 1980s, when I was still in college chasing my Electrical Engineering degree, I volunteered to be a stagehand for a 24-hour charity dance-a-thon at the college field house. Being a reasonably good musician and a novice sound person, I did not vie for the coveted front of house mix position, as every wannabe tech-head was already competing to hangout at FOH, and hoping that the hourly change in bands was not accompanied by band engineers. But I had a blast practicing changeovers onstage, calling out new patches on a new-fangled wireless intercom and keeping the less than reliable racks and stacks working around the clock.

But my biggest take home lesson on production was asking my 20 questions of the local volunteer soundco owner, who set up the rig and used us college kids as grunt labor. As part of setting up, I noticed a large metal and green painted fishing tackle box near the FOH location. When the owner opened up the tackle box, I gazed at the myriad of XLR adapters and other connectors with wonderment. From that day on, I vowed to assemble my own version of the green tackle box, to be prepared for any handy use. Now that my own tackle box is assembled, I want to infect you with the idea carrying your own tacklebox with similar just-in-case (JIC) adapters and miscellany.

THE BOX

Like my green tackle box mentor, I selected a large tackle box from Target Stores that had three cascading trays that could fit larger fishing lures, or, in my case, XLR barrel adapters, sharpies and other goodies. Today the fold-out tray tackle boxes are mostly out of style, but you can still find many used boxes at yard sales. I also looked for a tackle box with a large lower cavity to stow odd items like tape and hand tools, as they do not fit lure shapes. **Figure 40-1** shows an "as-is" picture of my tackle box. To convert my tackle box to a JIC box, I just wrote my name top-

Figure 40-1

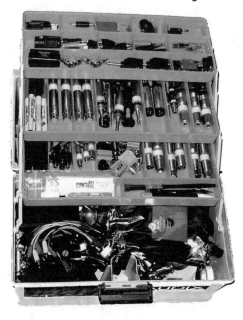

side and adorned the box with bumper stickers of products I use and radio stations I listen to.

My box is plastic, and is showing signs of age like cracks from things bouncing on it during transport. I can still get a few more years out of it, in a gaff tape purgatory, but it has seen a lot of inventory changeover in the last three decades of service.

TOYS IN THE ATTIC

When I first started filling the JIC box, my head was in the 1980s, and I was not totally compelled to wire everything with balanced patch cords. So I made a few missions to Radio Shack to stock up on both XLR to Phone transformer adapters, and other between connector adaptors.

Having seen many barrel adapters (XLR-male/female) in the green tackle box for attenuators, polarity flips and on/off switches, I bought a bunch of Shure blank barrels, and made my own essential XLR-through adapters. Armed with my *Audio Cyclopedia* balanced attenuator tables, I made my own 6 dB and 20 dB pad attenuators that have served me very nicely through the years. And simpler items like polarity flips and 2/3 pin short switch adapters have also proved handy. Every adapter and widget got labeled and brushed with clear fingernail polish so it wouldn't get misplaced or pilfered easily.

UNIQUE ITEMS

Besides adapters, Sharpies, hand tools like screwdrivers and pliers and many kinds of tapes, I was inspired to add other unique items to the JIC. One very handy item is a cheapy half-inch paint brush, useful for herding dust off of console work surfaces. Even with the best coverings and due diligence, consoles do collect dust, and a brush is much neater than your greasy fingers. Keen observers will note my collection of three-prong to two-prong grounding adapters. These are mainly for historical purposes, and only used to appease the hum gods if desired.

Lately my box well is filled with mic clips, homebrew Speakon to banana plug/jack adapters, spare Littlite bulbs, a Gerber multi-tool, a pocket digital multi-meter and my last gadget: a battery powered soldering iron. At six watts, I don't use it for large wire soldering, but it is more than handy on line-level cabling. Another handy item picked up at winter NAMM 2005 is a high-output clamp-on white LED "Bil-Lite" made by QED Inc. It was only $10, and has paid itself off many times over on briefcase gigs where no console lamp is available.

One item I have used sparely, but is still really handy to have, is a key receptacle lock-out. This is a red, two-prong receptacle that you can place over important plugs (like console power supplies) to prevent someone from firing up your rig, should you not be present between gigs. There is nothing as worrisome as walking into a venue, and seeing that some stranger (or musician) has powered up everything, making you wonder how many pops and thumps your speakers endured while they fumbled through the initialization sequence.

While I have mostly safe confines and security personnel at the gigs I do, the lockouts offer another level of peace of mind to keep the ambitious away.

YOUR TURN

If you have unique items in your JIC box, please write us at *FOH* Magazine — or visit the Wild and Woolly forums at www.fohonline.com — and share the wisdom of your unique JIC box items. While I can not carry big things in my box, little necessities like my Caig Labs De-oxit pen, my Fluke Volt-Alert pen and NL4 Speakon cable extenders (NL4MM) make handy case stuffers.

41 MONITOR MIXING BASICS

July 2006

Like it or not, monitor mixing tends to get a "waterboy" image compared to the front of house mix position. But a bad monitor mix situation can actually create a bad performance as performers suffer through less-than-ideal conditions. And interestingly enough, a bad FOH mix may not even be noticed by the performers if they're enjoying a great monitor mix that creates a good vibe onstage.

I want to share my mindset when monitor mixing, because one can achieve great pride doing this task, and not have to worry much about the prima donna working out at the FOH console. After all, you are working for the band, and your job is to deliver maximum satisfaction to them within the limits of the equipment. As a performer, though, I also look at monitor mixing as how I would want things done for me if I were performing on the stage. If you are a current or former musician, monitor mixing should be a position you should desire.

IN THE BEGINNING

Back when I started playing guitar with bands in the mid-1970s, we did not even have stage monitor speakers. Any clues on how our vocals blended or how the vocals and instruments mixed was found by hearing the echo off the back wall of the barroom. Being a curmudgeon of sorts, I do not hesitate to bring up that anecdote when I run into young musicians insisting on elaborate monitor mix setups when limited time or gear prevent the request from being honored. My band's first monitor system consisted of a 60-watt Bogen amplifier driving three homebrew single 12-inch woofer wedges (what tweeters?) to the performers, taken off a tap from the main mix.

Today we can have many mixes using dedicated monitor mixing consoles, high power amplifiers and monitor wedges of fine quality. And with enough mixes and input channels, mixing for personal monitors is finally becoming feasible for mid-level performers to implement. While I will not cover the additional skills on PM mixing, mixing for wedges covers many of the same tasks.

PREPARATION

When coming onto a monitor mixing gig, knowledge of the artists and the gear makes a big difference in the stress levels of the gig. If I don't have any knowledge of either I usually scout the venue, or call the providing soundco to find out what console and wedges they'll be using, as well as any other specifics like amps (bi-amp or passive) and number of mixes possible. If I have done the act before then I feel better about their needs, and can focus on lining up the gear. But if I don't know the performers then I will either want to advance with them for their monitoring needs, or at least query them as soon as possible before soundcheck.

Once I have determined the gear and its starting configuration, the first thing I want to do is zero out the console. Zeroing for me means cranking back the gains to at least the 10 o'clock positions, and checking the pads and polarity switches for normal settings. The monitor console channel strip EQs get flattened at first, unless there is an automatic tweak I do to certain instruments. A good example would be the lower-mid, wide-Q dip on the kick drum mic input.

Next, I zero the mix/aux feeds for each mix. I start with every feed completely off, then methodically build vocals-only mixes for each wedge mix. A typical example would be three downstage mixes labeled 1, 2 and 3 from stage-right, center and stage-left. Knowing which downstage vocal mics are on which channels, like 17, 18, 19 from stage-right to left, I push each mix/aux send to 2 o'clock as a starting point. Then I'll bring the other vocals into each mix at the 12 o'clock positions, so that each vocalist gets their vocal on top of their mix — just in case their egos get the best of them. Then I'll create a drum fill mix with 12 o'clock vocals, starting with a 10 o'clock kick mic setting.

The last zeroing item is to bring the mix masters to zero before firing up the EQs, amps and wedges. Then I'll check out all the remaining signal chains for sanity settings —things like hacked up graphic EQs and mis-set crossovers get attention before they get powered up. If I see lightly sloped EQs and proper cross-over settings I will assume all is well and go on to wedge positioning and powering up the amplifiers. If I have speaker processors on wedge mixes the sanity check generally goes deeper, but typically processor owners know what they are doing.

THE MOMENT OF TRUTH

By this time you should be feeling somewhat confident that everything is set to your desires and be ready for more confidence building. So get your amps powered up and get your talkback mic wired up and ready to go. Put talkback into each mix master and slowly bring up the fader chosen for each mix. You should get your voice through the desired wedges. Do this separately and note mentally the approximate fader levels on each mix. If there are flips in which wedge is getting what mix, or no talkback at all, then it is time to troubleshoot.

Once you have talkback signals confirmed in each mix, you can do two things. One would be to bring all the mixes up to nominal levels, then get an assistant to work each vocal mic up to proper channel strip gains and note any bad frequency response issues. Things like blown or buzzy tweeters and bad EQ settings can receive further attention. Once things sound normal, you can turn down all the mixes and proceed to ring out.

RING OUT

One mix at a time, go back to the nominal levels from before, and then slowly start pushing each mix to the point of ringing. Then it is just the tried-and-true method of identifying frequencies, notching them out 6 dB, and cranking up the fader for another ring or two. There is no sense doing a half dozen rings per mix, as the first couple will give you the most gain and the rest will only result in modest incremental dB gains. For me and my rig, I cheat — I ring out everything quickly, pull the faders off, and switch on the graphic EQ feedback detectors on my dbx iEQ31s. With six fixed filter notches, and six more floating notches, I don't hear a squeal from the wedges the rest of the night.

THE WAITING GAME

After ringing out, the rest of the night is waiting on the performers, adjusting mics and DI boxes and filling in their mix requests. Most musicians are reasonable people and only ask for a couple extra inputs beyond vocals. Beware of the ego-monsters with the "more me in the wedge" with their instrument. So far, my worst monitor mix night was a 100-watt-Marshall-stack-playing guitarist who doubled on a fiddle that was also plugged into the amp. He, of course, wanted more Marshall in his wedge while performing fiddle solos. You just have to cross your fingers, and hope there is a bit more feedback headroom between the amp mic and the wedge parked six feet in front of the mic.

During the show, be responsive to the performers and keep your focus on the show as best as possible. Do not go about making tweaks of large impact, or you will soon be persona non grata with the performers. If you hear some edges of feedback, you can PFL or AFL around mixes and channels and be proactive about notching out a suspect ring. And as I have mentioned several times over the years, you can monitor the PFL/AFL line with an RTA display to help find those "almost rings." From my trusty Neutrik Mini-lyzer to SMAART-Live software you can give your ears some help with your eyes.

42 TEACHING MIC TECHNIQUE

August 2006

As self-proclaimed audio experts, it our duty to learn the ins and outs of how microphone elements work, and how best to use them. Even though we are not Diva or Mister Ego vocalists, we should know how to work the mics. This is especially the case in house Of worship (HOW) situations where most of the speakers and singers are not professionals. So let's get on with teaching the teachers.

THE MIC FITS THE TASK

In my mic locker, I keep two sets of dynamic vocal mics. One set is the venerable Shure SM58 for those softer music acts and speaking engagements. Yes, these microphones were brand new when the Woodstock festival overtook upstate New York, and even Shure has their Beta series to improve upon this nearly 30-year-old design, but the SM58 mics wide cardioid pattern is very forgiving of bad mic technique. For the very loud bands, I break out the Audix OM-5s because of their super/hyper-cardioid pattern which only captures vocals from straight in. And of course you can pick your favorite modern tight pattern mics that suit your style and tastes.

SPEAKING INTO THE MIC

If you have new talent that is just supposed to be doing speech-making or mostly talking, tell the speaker to stay about "one fist away" from the windscreen. Eating the mic does not make sense for conversational work, and as most newbies cannot control their distance to the mic very well, the fast changes in volume due to distance is more aggravating to you and the audience. More than one fist away from the mic invites acoustic feedback — the console operator has diminished vocal intensity, has to turn up the volume, and the speaker leakage has a better chance of winning its battle with the talker's intensity.

SINGING INTO THE MIC

If the new person handling the mic is to sing, tell them to keep their lips partly touching the mic. Now I know this sounds unsanitary, but you have your canister of Wet Ones wipes in your mic locker to clean up the windscreens after the performance, right? Most mics are designed to handle this close proximity vocal sound pressure, and some people sing softer than when they speak, or intentionally sing softer as a way of connoting more emotion during parts of a song. This is just one reason why newbie singers should be eating the mic.

HANDLING THE MIC

Strongly encourage new singers or talkers to use a mic stand or podium mic. Most newbies do not have enough experience to carry a tune and try to play entertainer at the same time. The worst offense a singer can commit is to hold the mic towards the ground when not singing. This is especially nasty if floor-located stage monitors are employed. So please, break that habit by warning them in advance of what could happen. Keeping the mic on the mic clip when not in use, or holding the mic upright at their chest, is the recommended at-rest location.

The grip of the mic is also extremely important for best quality of sound — especially in light of rap singers who cup the mic's windscreen as a style. Most cardiod pattern mics need a clear area beneath the windscreen so that local sound waves can enter into side ports of the mic capsule and cancel out sound from anyplace other than the front of the mic. Please teach the newbie vocalists to take cues from the mic handgrip design and keep the fingers on the narrowest part of the mic grip. Fight the "cool look" grip with logic and truth, using such statements as "You will sound best with a lower grip".

DIVA GRIP — THE ICE CREAM CONE

Some of you may have noticed male and female lead vocalists using the "ice cream cone" mic technique, in which the mic is always held completely vertical and on the chin, so that the camera can pick up that million-dollar smile and expressions. While a nice way to hold a mic, it has two drawbacks. First, not every mic has a wide enough cardioid pattern to pick up vocal sound pressures without altering the frequency response greatly. The second drawback is that the singer must be a fairly powerful vocalist to cover the increased distance from the mouth's pressure zone.

HAVE COMPASSION

Finally, please have compassion for your newbie vocalists. They have a ton of things on their mind — dealing with stage fright, remembering stories and lyrics, their body language training/entertaining, not to mention your few words of wisdom on mic technique. Sometimes just a little eye contact and gesture to hold mic closer is what a new vocalist needs when in the midst of performance. Now you are the teacher…